Clinical Thermology

Subseries Thermotherapy

M. Gautherie (Ed.)

Methods of External Hyperthermic Heating

With Contributions by
J. W. Hand · K. Hynynen · P. N. Shrivastava · T. K. Saylor

With 121 Figures and 34 Tables

Springer-Verlag Berlin Heidelberg New York
London Paris Tokyo Hong Kong

Dr. Michel Gautherie
Laboratoire de Thermologie Biomédicale
Université Louis Pasteur
Institut National de la Santé
et de la Recherche Médicale
11, rue Humann
67085 Strasbourg Cedex, France

ISBN-13:978-3-642-74635-2   e-ISBN-13:978-3-642-74633-8
DOI: 10.1007/978-3-642-74633-8

Library of Congress Cataloging-in-Publication Data
Methods of external hyperthermic heating / M. Gautherie (ed.); with contributions by J. W. Hand ...
[et al.]. p.   cm. − (Clinical thermology. Subseries thermotherapy) Includes bibliographical references.
ISBN-13:978-3-642-74635-2 (U.S.)
1. Thermotherapy. I. Gautherie, Michel. II. Series.
RM865.M48   1990   615.8′32 − dc20   89-21981

Typesetting: K + V Fotosatz GmbH, Beerfelden

2127/3145-543210 − Printed on acid-free paper

Preface

The development of equipment capable of producing and monitoring safe, effective and predictable hyperthermia treatments represents a major challenge. The main problem associated with any heating technique is the need to adjust and control the distribution of absorbed power in the tissue during treatment. Power distribution is considered adequate only when tumor tissue can be maintained at the required hyperthermic levels while, at the same time, healthy tissue is not overheated. This problem is particularly crucial when external heating devices are used to produce hyperthermia. External hyperthermia refers to those methods which supply heat to tumor tissue in an external, noninvasive manner, as opposed to internal hyperthermia by which heat is supplied to tumor tissue in situ.

Until recently, most of the technical developments and clinical trials of thermotherapy for superficial and deep tumors have been based on electromagnetic systems. Presently, there is increasing interest in the use of ultrasound to accomplish these goals. Electromagnetic techniques of external thermotherapy include radiative, capacitive, and, to a lesser extent, inductive procedures. Recent designs for radiative applicators have incorporated microstrip structures. These have the advantage of being compact and lightweight compared with dielectrically loaded waveguide applicators. When using radiative applicators, proper control of power distribution can be achieved by scanning the applicator over the tissues or using arrays of simple applicators, such as annular phased arrays in which relative powers and phases are adjusted electronically. Capacitive electrodes have also been utilized extensively, based upon their capacity to deliver heat at depth. Control over power deposition, however, is difficult, and problems arise when thick layers of subcutaneous fat are present and when tissue heterogeneity leads to regions of high-current density and related hot spots.

A variety of ultrasound heating systems are being investigated. For the treatment of superficial tumors, single-plane transducers are useful, but multielement applicators offer greater control over the power deposition pattern. Recent designs for deep heating apply advanced ultrasound technologies such as multiple focussed transducers moved mechanically in such a way that the heating foci are scanned through the tumor volume, or phased arrays with electronic scanning allowing complex spot-focus scan paths for precise synthesis of heating patterns. In spite of significant limitations related to intervening bone and gas, ultrasound systems are suitable for external heating of superficial and deep tumors in a variety of anatomical sites.

Regardless of the method selected, quality control is of paramount importance for two reasons. The first is to evaluate the heating capabilities of the equipment; the second is to objectively compare the results of multicenter trials performed using different heating systems. Procedures, guidelines, and study criteria are presently being discussed by the European Society for Hyperthermic Oncology (ESHO), the North American Hyperthermia

Group (NAHG), and the Japanese Society of Hyperthermic Oncology (JSHO). The use of standard, tissue-equivalent phantoms has been recommended to investigate power deposition patterns as measured by the distribution of the so-called specific absorption rate (SAR).

The techniques for providing external hyperthermia have improved considerably during the past decade. Further progress is required, however, particularly to develop a precise concentration of hyperthermia within the target tumor volume. In this regard, the potentials of microwave and ultrasound heating devices appear to be greater than those of other modalities. The design and development of any thermotherapy equipment should always consider in parallel the heat-producing and the temperature-measuring systems in order to allow feed-back control of tumor hyperthermia.

Strasbourg, January 1990   M. GAUTHERIE

Contents

List of Contributors

J. W. HAND
Medical Research Council, Cyclotron Unit, Hammersmith Hospital, Ducane Road, London W12 0HS, Great Britain

K. HYNYNEN
Dept. Radiation Oncology, University of Arizona, Health Science Center, Tucson, Arizona 85724, USA

T. K. SAYLOR
Allegheny Singer Research Institute, 320, East North Avenue, Pittsburgh, PA 15212, USA

P. N. SHRIVASTAVA
Department of Radiation Oncology, University of Southern Carolina, School of Medicine, Los Angeles, CA 90033, USA

1 Biophysics and Technology of Electromagnetic Hyperthermia

J. W. HAND

1.1 Electromagnetic Fields and Tissues

1.1.1 Introduction

According to Licht (1965) in his excellent account of the history of the therapeutic use of heat, the earliest medical applications of electric current were developed during the 1830s and 1840s. They included the coagulation of blood in aneurysms and the destruction of fungoid growths. Shortly afterwards Sedillot (1853) reported using electric cautery to destroy tumours. Indeed, the use of cautery became widespread during the following decades, leading to the development of the galvanic knife (Boeckel 1873). These early techniques involved the application of dc currents directly to the tissues.

In the later decades of the nineteenth century scientists, including Thomson, Maxwell, Herz and Marconi, experimented with oscillating currents. Much of the pioneering work involving medical applications of radiofrequency currents was carried out by d'Arsonval at the Collège de France in Paris and a detailed account of his work has been given recently by Guy (1984). For example, he was the first to demonstrate a lack of physiological effect at frequencies above 10 kHz. He presented his data, including reference to the use of currents with frequencies greater than 1 MHz, to the Société de Biologie in Paris in 1891. In the following year he demonstrated that a high frequency current of 1 A could be passed through an assistant, himself and a filament lamp and that the only sensation he and his assistant experienced was that of warmth. Low frequency currents of such magnitude passing through the human body had usually proved fatal. In 1893 d'Arsonval introduced an induction system which, for the first time, permitted application of high frequency currents without the need for direct contact with the tissues. Soon afterwards he introduced the technique of capacitive coupling to transfer RF energy to the patient. In reviewing his own experiments d'Arsonval concluded that " ... high frequency currents penetrate profoundly into the body instead of accumulating on the surface". Meanwhile in 1891 Tesla, who had begun working with high frequency currents independently in North America, suggested that they might have medicinal value. In 1899 von Zeynek suggested that high frequency currents produced heat in tissues in the same way as in any other resistance. By the beginning of the present century the use of high frequency (several hundred kilohertz) currents became popular. In 1907 Nagelschmidt showed that high frequency currents produced heating at depth in tissues and referred to the process as "diathermy" (Nagelschmidt 1913). Treatment of rheumatism and rheumatic arthritis by high frequency currents was reported by Von Zeynek et al. (1908).

In early machines a spark discharge from electrodes which usually consisted of gas-filled glass tubes was played over the patient. The sparks, known as effluvia, could be up to 20 cm in length and were associated with electromagnetic fields of frequency 2-3 kHz. The later machines which were used in diathermy worked with spark gaps and the frequencies of these were generally in the range 0.6-1 MHz. The considerable damping of the fields required the patient to be very close to the source and the outputs of these machines were unpredictable. The use of thermionic valves as sources of undamped high frequency power was investigated during the early 1920s and the development of radio broadcasting led to the manufacture of increasingly better vacuum tube oscillators. Gosset et al. (1924) investigated the effects of 150-MHz currents upon plants tumours and in the same year Schereschewsy began his experiments with capacitively coupled short-wave fields. Tumours were implanted into mice and rats and were heated between a pair of electrodes. Regression was observed in 60%-70% of tumours treated in this way. General exposure of the tumour-bearing animals to the short-wave field did not affect tumour growth. Initially it was thought that the observed effects were due to some specific effect of frequencies in the range 20-80 MHz (Schereschewsky 1928) but later work (Christie and Loomis 1929; Schereschewsky 1933) showed that heat was the cytotoxic agent involved. In 1926 Schliephake began investigating the biological effects of short-wave condenser (electric) fields and later he applied the technique to treating local regions in patients. His first treatment was to himself when he applied the technique to a nasal boil.

In the late 1920s and early 1930s there was much interest in the possibility of using electromagnetic power to induce deeper heating both in animals and in man. In America the aim of early applications of the short-wave current in humans was to elevate body temperature. In 1927 Neymann and Osborne were first to use high frequency currents to induce fever in dogs and later in man (Neymann 1934). Other early reports on the use of short-wave currents to produce fever in man were those of Nagelschmidt (1928) and Carpenter and Page (1930). Early reports on the possible use of diathermy to heat lung tissue include Binger and Christie (1927) and Christie et al. (1928).

The next few years saw the development of the short-wave diathermy technique (Neymann and Osborne 1929) in which fenestrated electrodes were used with a generator capable of delivering up to 750 W at frequencies in the range 0.5–1.5 MHz. One electrode was used to cover the patient's back and either one or two electrodes covered his chest and abdomen. Conductive jelly was used between the electrodes and the patient's skin to ensure good contact but many difficulties were encountered in implementing the technique safely in the clinic (Neymann 1938). In 1928 several reports of the use of machines operating at 50–100 MHz were published (Hosmer 1928; Neymann and Osborne 1929; Soiland 1928). In a detailed account of the use of thermal therapy during the 1930s, Neymann (1938) describes the early machines, known as radiotherms, used to induce an elevated body temperature in patients:

The patient lies in the treatment bag (which provided thermal insulation) on interlaced asbestos tapes stretched across a wooden frame. He rests on his back and the plates are placed at each side so that the waves oscillate through the body from one side to the other ... In this manner febrile temperatures can be produced at will ... Restless and violent patients twist around in the treatment bag and cross their arms and legs and are thus likely to be burned. Any peculiar position of the patient lying within the electromagnetic field which concentrates the power into a small area, such as touching two index fingers together, will intensify the heat at that point and finaly result in a burn. ... small burns under the armpits and wherever large amounts of perspiration gather are so frequent that they are a constant annoyance to the patient and the operator.

Clearly, the machine had a number of drawbacks:

There are many objections to the practical use of the apparatus. First, it is difficult to raise the temperature quickly and with precision when excitable patients are being treated. Second,

much heat is lost while lifting the patient out of the radiotherm to his bed. Third, the machine frequently catches fire.

Detailed accounts of the wide interest in the use of thermal therapy in the 1930s are given by Schliephake (1935) and Neymann (1938). Many short-wave machines appeared on the market and unsubstantiated and extravagant claims, including frequency-specific effects, were made by their manufacturers. Comparative tests and evaluations of various machines were carried out under the auspices of the Council on Physical Therapy of the American Medical Association (Council on Physical Therapy 1934). The results indicated that increases in temperature induced in the viscera of dogs were only about 0.5 °C. Furthermore, it was shown that the technique of electromagnetic induction in which the high frequency current was passed through a flexible cable wrapped around or about the region to be treated was superior to techniques employing capacitive electrodes. At the same time results of other studies failed to support claims for specific biological effects of high frequency currents (Eidinow 1935; Mortimer and Osborne 1935).

The therapeutic applications of RF power kept in step with technical developments. Denier (1936) in France was perhaps the first to use microwave (375 MHz) therapy. After World War II high power magnetrons became generally available and microwave (3000 MHz) techniques were developed (Krusen et al. 1947; Leden et al. 1947). Many claims, including frequency-specific effects, were made for the techniques. Often these arose from the failure to recognise the importance of careful dosimetry, an area pioneered earlier by Mittlemann et al. (1941), who measured changes in temperature and related them to absorbed energy. In the 1950s and 1960s the exensive work of Schwan (1957, 1959) and Lehmann and his colleagues (1962, 1965; Guy and Lehmann 1966) provided a deeper understanding of microwave diathermy and pointed to the advantage of using a frequency somewhat lower than 2450 MHz, a frequency which had been adopted for industrial, scientific and medical purposes.

During the 1970s and 1980s there has been a resurgence of interest in all aspects of hyperthermia research. On the technical side there are now improved applicators capable of heating superficial tumours. In the late 1970s development of the first electromagnetic applicators capable of depositing energy deep in the body began and this area remains one of activity today.

The field of clinical hyperthermia has matured in the past few years. In parallel with the increasing number of scientists and physicians showing interest in hyperthermia, there is an air of realistic optimism as we come to terms with the problems involved. Controlled clinical evaluations of hyperthermia are underway and it is essential that these and future studies should be based on the best practices available. To achieve this a full understanding of the devices used will be necessary.

1.1.2 Overview of the Chapter

A short discussion of electromagnetic fields, their propagation and interaction with biological tissues completes Sect. 1.1. Aspects of electromagnetic dosimetry relevant to hyperthermia using phantom materials and computer models are discussed in Sect. 1.2. Section 1.3 gives a overview of electromagnetic heating techniques. Details of applicators for local hyperthermia are discussed in Sect. 1.4 and applicators for regional hyperthermia in Sect. 1.5. Finally a summary of the biological effects of electromagnetic waves and details of exposure guidelines are given in Sect. 1.6.

1.1.3 Electric and Magnetic Fields

1.1.3.1 Time-Invariant Fields

There are two forms of electric charge − positive and negative. The force $\mathbf{F}_e$ between two charged spherical bodies whose radii are very small compared with their separation r and which are remote from other dielectric media is

$$\mathbf{F}_e = \frac{1}{4\pi\varepsilon} \frac{q_1 q_2}{r^2} \mathbf{r} \tag{1.1}$$

where q_1 and q_2 are the charges on the bodies and ε is the permittivity of the medium in which they are located. For free space $\varepsilon = \varepsilon_0 = 8.854 \times 10^{-12}\,\mathrm{F\,m^{-1}}$. Equation (1.1) is known as Coulomb's law of electrostatic force. Throughout this chapter **bold** type is used to indicate vector quantities.

If the charge q_2 is distributed over a region of space rather than being concentrated at a single point then the force experienced by q_1 is the vector sum of the forces due to all elemental charges dq_n which make up q_2. $\mathbf{F}_e$ becomes

$$\mathbf{F}_e = \frac{1}{4\pi\varepsilon} \sum_n \frac{q_1 dq_n}{r_n^2} \mathbf{r_n} \tag{1.2}$$

where r_n is the distance between q_1 and dq_n and $\mathbf{r_n}$ is the unit vector along the line joining them. If a test charge q is placed at some point P in the vicinity of the charge q_2 or charge distribution $q_2 = \sum q_n$ and the presence of q does not affect the neighbouring charge(s) (i.e. $q \to 0$), then the force per unit charge at P is

$$\mathbf{E} = \frac{\mathbf{F}_e}{q} = \frac{1}{4\pi\varepsilon} \sum_n \frac{dq_n}{r_n^2} \mathbf{r_n} \tag{1.3}$$

$\mathbf{E}$ is known as the electric field. If the charge distribution can be expressed in terms of a charge density $\varrho(\mathbf{r})$ then Eq. (1.3) becomes

$$\mathbf{E} = \frac{1}{4\pi\varepsilon} \int_v \frac{\varrho(\mathbf{r})\mathbf{r}}{r^2} dv \tag{1.4}$$

where dv is an elemental volume located at position $\mathbf{r}$. Equation (1.4) can be written in terms of the gradient of $(1/r)$ to give

$$\mathbf{E} = \frac{1}{4\pi\varepsilon} \int_v \varrho(\mathbf{r}) \nabla(1/r) dv \tag{1.5}$$

Since $\nabla \times \nabla \psi = 0$ for any vector which can be expressed as a gradient, it follows that

$$\nabla \times \mathbf{E} = 0 \tag{1.6}$$

which implies that any static electric field is conservative.

The electric field flux through an elementary area dS is defined as the product of the area and the component of electric field normal to that elementary area. Thus the total flux through some general surface is

$$\Phi_e = \int_S \mathbf{E} \cdot \mathbf{dS} \tag{1.7}$$

where $\mathbf{dS}$ is a vector with magnitude dS and direction along the outward normal to that element of surface. If $\mathbf{dS}$ is located at $\mathbf{r}$ with respect to a single charge q then

$$\int_S \mathbf{E} \cdot \mathbf{dS} = \int_S \frac{1}{4\pi\varepsilon} \frac{q}{r^2} \mathbf{r} \cdot \mathbf{dS} = \frac{q}{4\pi\varepsilon} \int_S d\Omega \tag{1.8}$$

where $d\Omega$ is the solid angle subtended at q by dS. The integral of the solid angle over a closed surface which includes the location of q is 4π and so

$$\int_S \mathbf{E} \cdot \mathbf{dS} = \frac{q}{\varepsilon} \tag{1.9}$$

When the charge distribution can be represented by a continuous charge density ϱ, Eq. (1.9) becomes

$$\int_S \mathbf{E} \cdot \mathbf{dS} = \frac{1}{\varepsilon} \int_v \varrho \, dv \tag{1.10}$$

This is the integral form of Gauss's equation which states that the total electric field flux passing out through a closed surface is equal to the total charge enclosed within the surface divided by ε. The differential form of this law may be obtained by taking Eq. (1.10) together with the divergence theorem

$$\int_S \mathbf{E} \cdot \mathbf{dS} = \int_v \nabla \cdot \mathbf{E} \, dv \tag{1.11}$$

to give

$$\int_v \left(\nabla \cdot \mathbf{E} - \frac{\varrho}{\varepsilon} \right) dv = 0 \quad \text{or} \quad \nabla \cdot \mathbf{E} = \frac{\varrho}{\varepsilon} \tag{1.12}$$

which shows that the electric field flux out per unit volume at a point is proportional to the charge density at that point.

To extend the discussion to magnetic fields we must consider moving charges in which case an additional force $\mathbf{F}_m$ is exerted on the charge. A vector field quantity known as the magnetic flux density $\mathbf{B}$ is defined by the equation

$$\mathbf{F}_m = q(\mathbf{v} \times \mathbf{B}) \tag{1.13}$$

where $\mathbf{v}$ is the velocity of q. The vector product in Eq. (1.13) implies that $\mathbf{F}_m$ is normal to both $\mathbf{v}$ and $\mathbf{B}$. The unit of $\mathbf{B}$ is the Tesla (formerly Wb m^{-2}).
If $\mathbf{dl}$ is an element of length of a wire carrying I_1, r is the separation between $\mathbf{dl}$ and a point of observation P and $\mathbf{r}$ is the unit vector directed from $\mathbf{dl}$ to P the elemental magnetic flux density $\mathbf{dB}$ at P is

$$\mathbf{dB} = \frac{\mu_0}{4\pi} I_1 \frac{\mathbf{dl}_1 \times \mathbf{r}}{r^2} \tag{1.14}$$

and the total magnetic flux density $\mathbf{B}$ at P due to the complete current circuit is

$$\mathbf{B} = \frac{\mu_0}{4\pi} I_1 \int_1 \frac{\mathbf{dl}_1 \times \mathbf{r}}{r^2} \tag{1.15}$$

This relationship is known as the Biot-Savart law. The parameter $\mu_0 = 4\pi \times 10^{-7} \, \text{H m}^{-1}$ is known as the permeability of free space. A very useful relationship, known as Ampere's circuital law, is

$$\oint_1 \mathbf{B} \cdot \mathbf{dl} = \frac{\mu_0}{4\pi} I \tag{1.16}$$

i.e. the line integral of $\mathbf{B}$ about any closed path is equal to $\mu_0/4\pi$ times the direct current I enclosed by that path.

1.1.3.2 Time-Varying Fields

Time-varying electric and magnetic fields are related to each other and to their sources, charge and current density, by the set of equations known as Maxwell's equations. The integral forms of Maxwell's equations in free space are

$$(I) \quad \int_S \mathbf{E} \cdot \mathbf{dS} = \frac{1}{\varepsilon_0} \int_v \varrho \, dv \tag{1.17}$$

which is Gauss's law [see Eq. (1.10)].

$$(II) \quad \oint_1 \mathbf{E} \cdot \mathbf{dl} = -\int_S \frac{\partial \mathbf{B}}{\partial t} \cdot \mathbf{dS} \tag{1.18}$$

is Faraday's law, which states that the electromotive force (emf) around any closed loop is equal to the time rate of change of the magnetic flux cutting that loop.

$$(III) \quad \int_S \mathbf{B} \cdot \mathbf{dS} = 0 \tag{1.19}$$

which states that the total magnetic field flux passing through any closed surface is zero, implying that magnetic field lines always form closed loops.

$$(IV) \quad \oint_1 \mathbf{B} \cdot \mathbf{dl} = \mu_0 \int_S \mathbf{J} \cdot \mathbf{dS} + \varepsilon_0 \mu_0 \int_S \frac{\partial \mathbf{E}}{\partial t} \cdot \mathbf{dS} \tag{1.20}$$

is Ampere's circuital law. $\mathbf{J}$ (A m^{-2}) is the current density. The first term on the right-hand side of Eq. (1.20) relates the magnetic field to the steady current whilst the second term involving the displacement current produced by the time rate of change of magnetic field is included to ensure that charge is conserved. The equation which must be satisfied for conservation of charge is

$$\int_S \mathbf{J} \cdot \mathbf{dS} = -\int_v \frac{\partial \varrho}{\partial t} \, dv \tag{1.21}$$

i.e. that the total current passing out through any closed surface is equal to the total change in charge within the surface.

The integral forms of Maxwell's equations are usually easier to explain in terms of the experimental laws from which they have been obtained but the differential forms are usually more convenient to handle mathematically. The differential forms are

(I) $\quad \nabla \cdot \mathbf{E} = \dfrac{\varrho}{\varepsilon_0}$ \hfill (1.22)

which shows that charge density is a source of electric field and that electric field lines begin and end on charges.

(II) $\quad \nabla \times \mathbf{E} = -\dfrac{\partial \mathbf{B}}{\partial t}$ \hfill (1.23)

which shows that the time rate of change of magnetic field is a curl-type source of electric field.

(III) $\nabla \cdot \mathbf{B} = 0$ \hfill (1.24)

which shows that **B** fields have zero divergence and so magnetic field lines are always closed.

(IV) $\nabla \times \mathbf{B} = \mu_0 \left(\mathbf{J} + \varepsilon_0 \dfrac{\partial \mathbf{E}}{\partial t} \right)$ \hfill (1.25)

which indicates that both current density and the time rate of change of electric field are curl-type sources of magnetic field.

Assuming a sinusoidal time dependence of $\mathbf{E}, \mathbf{B}, \mathbf{J}$ and ϱ, we can separate the temporal and spatial dependencies to obtain

$\mathbf{E}(x, y, z, t) = \bar{\mathbf{E}}(x, y, z) \exp [j \omega t]$

$\mathbf{J}(x, y, z, t) = \bar{\mathbf{J}}(x, y, z) \exp [j \omega t]$

$\mathbf{B} = \bar{\mathbf{B}}(x, y, z) \exp [j \omega t]$

$\varrho = \bar{\varrho}(x, y, z) \exp [j \omega t]$

Substituting these expressions into Eqs. (1.22 – 1.25) gives

$\nabla \cdot \bar{\mathbf{E}} = \dfrac{\bar{\varrho}}{\varepsilon_0}$ \hfill (1.26)

$\nabla \times \bar{\mathbf{E}} = -j \omega \bar{\mathbf{B}}$ \hfill (1.27)

$\nabla \cdot \bar{\mathbf{B}} = 0$ \hfill (1.28)

$\nabla \times \bar{\mathbf{B}} / \mu_0 = \bar{\mathbf{J}} + j \omega \varepsilon_0 \bar{\mathbf{E}}$ \hfill (1.29)

$\bar{\mathbf{E}}$ and $\bar{\mathbf{B}}$ are found from the solutions of Eqs. (1.26 – 1.29) and the real time forms of the fields are obtained by restoring the time dependence, i.e.

$\mathbf{E}(x, y, z, t) = \mathrm{Re}\{\bar{\mathbf{E}} \exp [j \omega t]\}$ and
$\mathbf{B}(x, y, z, t) = \mathrm{Re}\{\bar{\mathbf{B}} \exp [j \omega t]\}$

where Re indicates the real part of the complex quantity.

Equations (1.27) and (1.29) show that electric and magnetic effects are distinct phenomena when $d/dt = 0$. Even if $d/dt \neq 0$, the approximation that electric and magnetic effects can be considered separately is a good one as long as the time derivatives in Maxwell's equations remain insignificant. In practice this requires that the dimensions of the problem under consideration must be small with respect to the wavelength. Such fields are referred to as being quasi-static.

1.1.4 Electrical Properties of Biological Materials

In general, currents which are produced in materials by the interaction with **E** and **B** fields may be considered in terms of three mechanisms. The first of these is the drift of free conduction charges in the material in response to the applied **E** field. This drift is the resultant of movement due to field forces superimposed on the random motion of the charges due to their thermal energy. This conduction current is $\sigma \mathbf{E}$, where σ is the conductivity. The SI units of σ are $\mathrm{S\,m^{-1}}$ although a unit 100 times larger, $\mathrm{mho\,cm^{-1}}$, is often found in the literature.

The second mechanism is dielectric polarisation, which may arise when the positive and negative charges of a molecule ar displaced slightly from their equilibrium positions due to the applied **E** field. This displacement, which gives rise to a polarisation charge, is opposed by attractive forces between the negative and positive charges. A second type of polarisation arises from the fact that some molecules (polar molecules) possess permanent electric dipoles. In the absence of an applied field, these dipoles are randomly orientated due to thermal motions but when an **E** field is present the dipoles align with the applied field. The extent of this alignment depends upon the strength of the field and it is opposed by the random thermal motion of the molecules and mutual interactions between molecules. Both polarisation effects are taken into account in the dimensionless quantity known as the electric susceptibility, χ. For field strengths relevant to electromagnetically induced hyperthermia, the polarisation field and the polarisation current are given by $\varepsilon_0 \chi \mathbf{E}$ and $\varepsilon_0 \chi j \omega \mathbf{E}$, respectively.

The third general mechanism is known as magnetisation and involves the alignment of magnetic dipoles in a material in opposition to their random thermal motion. However, this type of interaction is insignificant in biological materials and will not be discussed here. It is useful to define

$$\varepsilon = \varepsilon_0 (1 + \chi + \sigma / j \omega \varepsilon_0) = \varepsilon_0 (\varepsilon' - j\varepsilon'') \tag{1.30}$$

ε is known as the complex permittivity of the material and ε' and ε'' are its real and imaginary parts, respectively. The additional polarisation charge is accounted for by ε' whilst the effect of the conduction of free charges is described by ε''. ε'' determines how much power is absorbed from the electric field and the term "lossy" is applied to materials which absorb significant power. A related parameter is the loss tangent, which is defined as $\tan \delta = \varepsilon''/\varepsilon'$. The relationship between the ε'', σ and the angular frequency of the field $\omega (= 2\pi f)$ is

$$\varepsilon'' = \frac{\sigma}{\omega \varepsilon_0} \tag{1.31}$$

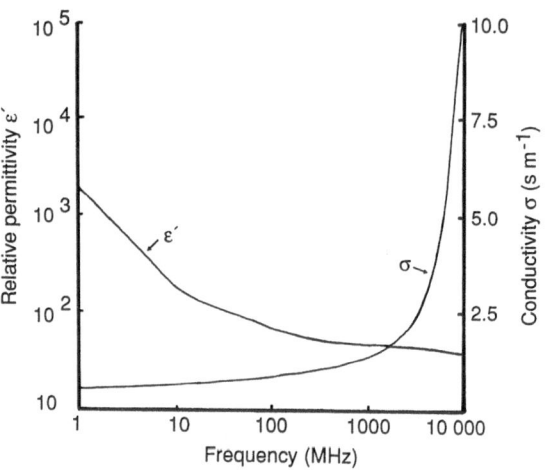

Fig. 1.1. Relative permittivity ε' and conductivity $\sigma (\text{S m}^{-1})$ of high water content tissue as functions of frequency

Both ε' and ε'' vary with the frequency of the applied field (see Fig. 1.1) and so are said to be dispersive. At frequencies lower than about 100 MHz, cell membranes, which have a capacitance of approximately $1 \mu\text{F cm}^{-2}$, act as thin layers of high impedance and restrict the flow of current to extracellular regions. Variations of ε' and σ in this frequency range depend upon the cellular structure of the tissue. At low frequencies the capacitance of the tissue and its dielectric constant are high since there is sufficient time to charge and discharge the cell membranes during each

Table 1.1. Relative permittivity ε' of selected tissues

f (MHz)	13	27	100	200	300	433	915	2450
High water content tissues[a]	160	113	72	57	54	53	51	47
Low water content tissues[a]		20	7.5	6.0	5.7	5.6	5.6	5.5
Blood[b]	200		73		63	63	60	58
Brain[c]	240[e]		70-83					
Eye lens[d]	100[e]	65[f]	48	40		34[g]	31[h]	30[i]
Heart muscle[d]				61		64[g]	53[h]	
Kidney[d]	220[e]	200[f]	90	62		54[g]	53[h]	51[i]
Liver[d]	143[e]	137[f]	78	53		48[g]	46[h]	
Lung[b]				35	36	36	35	32

[a] Johnson and Guy (1972)
[b] Iskander (1982)
[c] Durney (1987)
[d] Lin (1986)
[e] f = 10 MHz
[f] f = 25 MHz
[g] f = 400 MHz
[h] f = 1000 MHz
[i] f = 2500 MHz

Table 1.2. Conductivity $\sigma\,(\mathrm{S\,m^{-1}})$ for selected tissues

f (MHz)	13	27	100	200	300	433	915	2450
High water content tissues[a]	0.4	0.63	0.6	0.89	1.0	1.18	1.28	2.17
Low water content tissues[a]		0.011 – 0.043	0.019 – 0.076	0.026 – 0.094	0.031 – 0.110	0.038 – 0.118	0.056 – 0.147	0.096 – 0.213
Blood[b]	1.1		1.2		1.2	1.2	1.4	2.13
Brain[c]	0.222 – 0.333[e]		0.435 – 0.625					
Eye lens[d]	0.38[e]	0.4[f]	0.4	0.4		0.4[g]	0.5[h]	1.1[i]
Heart muscle[d]				0.96		1.09[g]	1.19[h]	
Kidney[d]	0.88[e]	1.0[f]	1.0	1.11		1.18[g]	1.23[h]	2.28[i]
Liver[d]	0.88[e]	0.51[f]	0.59	0.79		0.86[g]	0.98[h]	
Lung[b]	0.67[e]		0.71	0.63		0.71[g]	0.73[h]	

[a] Johnson and Guy (1972)
[b] Iskander (1982)
[c] Durney (1987)
[d] Lin (1986)
[e] f = 10 MHz
[f] f = 25 MHz
[g] f = 400 MHz
[h] f = 1000 MHz
[i] f = 2500 MHz

cycle. As the frequency increases the capacitive reactance of the membranes decreases, resulting in an increase in the conductivity of the tissue. The variation in ε' over this frequency range arises from the incomplete charging and discharging of the cell membranes during each cycle. At frequencies greater than about 100 MHz the capacitive reactance of the cell membranes is so small that they are short-circuited and the electrical properties of the tissue are then determined by the water, salt and protein content. Biological molecules may be considered as "dielectric holes" in the electrolyte. The slight frequency dependence of ε' and σ over the range 100 – 1000 MHz may be attributed to the rotation of these large polar molecules. The rapid increase in σ at frequencies above 1000 MHz and decrease in ε' above 3000 MHz result from the polar properties of the water molecules in the tissue.

Because of the low electrolyte content of fat and bone, the values of ε' and σ for these tissues are approximately one order of magnitude lower than those for tissues of high water content. Large variations in dielectric properties may be observed from one "dry" tissue to another since relatively small variations in the content of both free water and water in the immediate environment of the biological macromolecules can produce significant changes in ε' and σ.

Relative permittivity ε' and conductivity σ for selected tissues are listed in Tables 1.1 and 1.2.

1.1.5 Wave Propagation in Tissues

By analogy with Eqs. (1.26 – 1.29) Maxwell's equations in an unbounded region of tissue may be written

$$\nabla \cdot \bar{\mathbf{E}} = \frac{\bar{\varrho}}{\varepsilon} \tag{1.32}$$

$$\nabla \times \bar{\mathbf{E}} = -j\omega\mu_0\tilde{\mathbf{H}} \tag{1.33}$$

$$\nabla \cdot \tilde{\mathbf{H}} = 0 \tag{1.34}$$

$$\nabla \times \tilde{\mathbf{H}} = \bar{\mathbf{J}} + j\omega\varepsilon_0\bar{\mathbf{E}} \tag{1.35}$$

where ε_0 has been replaced by $\varepsilon = \varepsilon_0(\varepsilon' - j\sigma/\omega\varepsilon_0)$ and $\mathbf{B}$ has been replaced by $\mu_0\mathbf{H}, \mathbf{H}$ $(\mathrm{A\,m^{-1}})$ being the magnetic field intensity. A time dependency of $\exp[j\omega t]$ is assumed. By substituting the curl of Eq. (1.33) into Eq. (1.35), using the vector identity $\nabla \times \nabla \times \mathbf{E} = \nabla(\nabla \cdot \mathbf{E}) - \nabla^2\mathbf{E}$ and assuming that there are no free electric charges or currents ($\varrho = 0$ and $\mathbf{J} = 0$) in the region of interest we obtain

$$\nabla^2\mathbf{E} + \omega^2\mu_0\varepsilon\mathbf{E} = 0 \tag{1.36}$$

The simplest solution to Eq. (1.36) is a uniform plane wave which may be expressed as

$$\bar{\mathbf{E}}_{\mathbf{x}}(\mathbf{z}) = \bar{\mathbf{E}}_1 \exp[-\gamma z] + \bar{\mathbf{E}}_2 \exp[\gamma z] \tag{1.37}$$

where $\bar{E}_1$ and $\bar{E}_2$ are constants to be determined by boundary conditions and γ (m^{-1}), called the propagation constant, is

$$\gamma = j\omega\mu_0\varepsilon_0(\varepsilon' - j\sigma/\omega\varepsilon_0) \qquad (1.38)$$

γ is a complex quantity and can be separated into real and imaginary parts β and α, respectively. α (m^{-1}) is termed the attenuation constant and is

$$\alpha = \frac{\omega(\mu_0\varepsilon_0\varepsilon')^{0.5}}{2^{0.5}}\left\{\left[1 + \left(\frac{\sigma}{\omega\varepsilon_0\varepsilon'}\right)^2\right]^{0.5} - 1\right\}^{0.5} \qquad (1.39)$$

and β (radian m^{-1}), the phase constant, is

$$\beta = \frac{\omega(\mu_0\varepsilon_0\varepsilon')^{0.5}}{2^{0.5}}\left\{\left[1 + \left(\frac{\sigma}{\omega\varepsilon_0\varepsilon'}\right)^2\right]^{0.5} + 1\right\}^{0.5} \qquad (1.40)$$

Rewriting Eq. (1.37) in terms of α and β and incorporating the exp [jωt] time dependence gives

$$\mathbf{E} = \text{Re}\{\bar{\mathbf{E}}(z)\exp[j\omega t]\} = \mathbf{E}_1\exp[-\alpha z]\cos(\omega t - \beta z) \qquad (1.41)$$

as a particular solution of the time-dependent vector wave equation. Equation (1.41) describes a linearly polarised plane wave propagating in the z-direction with amplitude E_1 directed along x. A similar solution may be derived for the magnetic field $\mathbf{H}$ which is directed along y. Thus both $\mathbf{E}$ and $\mathbf{H}$ are perpendicular to z, the direction of propagation. The term exp [$-\alpha z$] indicates an exponential decay due to energy given up to the medium.

The ratio of the electric field to the magnetic field is known as the intrinsic wave impedance η.

$$\eta = \left[\frac{\mu_0}{\varepsilon_0\{\varepsilon' - j(\sigma/\omega\varepsilon_0)\}}\right]^{0.5} = \frac{\eta_0(\cos\theta + j\sin\theta)}{[1 + (\sigma/\omega\varepsilon'\varepsilon_0)^2]^{0.25}} \qquad (1.42)$$

where $\eta_0 = (\mu_0/\varepsilon_0)^{0.5} = 120\pi$ ohm and $\theta = 0.5\tan^{-1}(\sigma/\omega\varepsilon'\varepsilon_0)$. Values of η at several frequencies for tissue with high water content are given in Table 1.3. For a given frequency f, the wavelength λ is shorter in the tissue than in free space. It is

$$\lambda = \frac{2\pi}{\beta} = \frac{2^{0.5}}{f(\mu_0\varepsilon_0\varepsilon')^{0.5}\{[1 + (\sigma/\omega\varepsilon_0\varepsilon')^2]^{0.5} + 1\}^{0.5}} \qquad (1.43)$$

Wavelengths in tissue with high water content at several frequencies are given in Table 1.3.

1.1.6 Power Absorption

The time rate at which energy associated with the electromagnetic field in a volume V changes is given by

$$-\frac{d}{dt}(W_H + W_E) - \int_S \mathbf{E}\times\mathbf{H}\cdot\mathbf{ds} = \int_v \mathbf{E}\cdot\mathbf{J}\,dv \qquad (1.44)$$

where

$$W_H = \int_v \frac{\mu_0\mathbf{H}\cdot\mathbf{H}}{2}\,dv \quad \text{and} \quad W_E = \int_v \frac{\varepsilon\mathbf{E}\cdot\mathbf{E}}{2}\,dv$$

are the instantaneous quantities of energy stored in the magnetic and electric fields, respectively, and $\mathbf{dS}$ is a vector with magnitude dS (an elemental area of the surface enclosing V) and direction along the outward normal to that element of surface. $\mathbf{E}\times\mathbf{H}$, known as the Poynting vector, represents the instantaneous power density at a point and $\int_S \mathbf{E}\times\mathbf{H}\cdot\mathbf{dS}$ is the inward rate of energy flow through the surface of the volume.

$\int_v \mathbf{E}\cdot\mathbf{J}\,dv$ represents the instantaneous power dissipated within the volume. Since $\mathbf{J} = \sigma\mathbf{E}$, this term represents heat producing ohmic losses.

From a measurement point of view, time-averaged quantities are more useful than the instantaneous quantities in Eq. (1.44). The average Poynting vector is $\mathbf{E}\times\mathbf{H}^*$(W m^{-2}) (where $\mathbf{H}^*$ is the complex conjugate of $\mathbf{H}$). Thus

$$\int_V \text{Re}(\mathbf{J}^*\cdot\mathbf{E})\,dv = -\int_S \text{Re}(\mathbf{E}\times\mathbf{H}^*)\cdot\mathbf{dS} \qquad (1.45)$$

or the time average power dissipated in V is equal to

Table 1.3. Intrinsic wave impedance η and wavelength λ in tissue with high water content assuming the dielectric properties given in Tables 1.1 and 1.2

f (MHz)	100	200	300	433	915	2450
η (ohm)	$24 + j15.5$	$32 + j17.7$	$32 + j17.6$	$41.3 + j16.2$	$48.7 + j11.3$	$52.8 + j8.7$
λ (mm)	270	166	118	88	45	18

the time average power passing into V through the surface S. Using $\mathbf{J} = \sigma\mathbf{E}$ we find that the average power transferred to the material in the elemental volume dv is

$$P = \frac{1}{2}\,\sigma\,|E|^2 \tag{1.46}$$

where E is the magnitude of the electric field.

The specific absorption rate (SAR) (i.e. mass normalised energy absorption rate) is defined (NCRP 1981) as the time derivative of the incremental energy (dW) absorbed by (dissipated in) an incremental mass (dm) contained in a volume element (dv) of a given density (ϱ)

$$SAR = \frac{d}{dt}\left[\frac{dW}{dm}\right] = \frac{d}{dt}\left[\frac{1}{\varrho}\frac{dW}{dv}\right] = \frac{\sigma}{2\varrho}\,|E|^2 \tag{1.47}$$

The units of SAR are $W\,kg^{-1}$.

1.1.7 Boundary Conditions

When wave propagation occurs through more than one dielectric medium, Maxwell's equations must be satisfied in each medium and at the boundaries between media.

The conditions which must be satisfied at any boundary between two media, a and b, are

$$(\mathbf{E_a} - \mathbf{E_b}) \times \mathbf{n} = 0 \tag{1.48}$$

$$(\mathbf{D_a} - \mathbf{D_b}) \cdot \mathbf{n} = (\varepsilon_a \mathbf{E_a} - \varepsilon_b \mathbf{E_b}) \cdot \mathbf{n} = \varrho \tag{1.49}$$

$$(\mathbf{H_a} - \mathbf{H_b}) \times \mathbf{n} = \mathbf{J} \tag{1.50}$$

$$(\mathbf{B_a} - \mathbf{B_b}) \cdot \mathbf{n} = (\mu_a \mathbf{H_a} - \mu_b \mathbf{H_b}) \cdot \mathbf{n} = 0 \tag{1.51}$$

where $\mathbf{n}$ is a unit vector in the direction normal to the boundary.

When there are no surface charges or currents $(\varrho = 0$ and $J = 0)$ then the components of the electric and magnetic fields tangential to the boundary are equal

$$\mathbf{E_a^{tang}} = \mathbf{E_b^{tang}} \quad \text{and} \quad \mathbf{H_a^{tang}} = \mathbf{H_b^{tang}} \tag{1.52}$$

and the components of $\mathbf{D}$ and $\mathbf{B}$ normal to the boundary are equal. Since $\mu_a = \mu_b = \mu_0$ for tissues we have

$$\varepsilon_a \mathbf{E_a^{perp}} = \varepsilon_b \mathbf{E_b^{perp}} \quad \text{and} \quad \mathbf{H_a^{perp}} = \mathbf{H_b^{perp}} \tag{1.53}$$

These conditions must be satisfied regardless of the geometry of the boundary. For example, when a plane wave is incident along the normal to the plane boundary between two tissues, a transmitted wave will propagate in the second tissue and a reflected wave will propagate back into the first tissue, both waves being normal to the boundary. The reflection coefficient Γ is

$$\Gamma = \frac{\eta_b - \eta_a}{\eta_b + \eta_a} = \frac{\varepsilon_a^{0.5} - \varepsilon_b^{0.5}}{\varepsilon_a^{0.5} + \varepsilon_b^{0.5}} \tag{1.54}$$

where η_a, η_b are the intrinsic wave impedances of the two tissues. For a discussion of the more general case of reflection and refraction at boundaries for oblique incidence the reader is referred a standard text on electromagnetic theory (e.g. Stratton 1941).

1.1.8 Summary

1. Time-invariant electric and magnetic fields were introduced followed by a summary of Maxwell's equations describing time-varying relationships between the fields.
2. A summary of the electrical properties of biological tissues was given and the concept of complex permittivity was introduced.
3. Wave propagation in tissue was discussed in terms of Maxwell's equations for lossy dielectric media. A plane wave propagating in an unbounded region of tissue was used to introduce the attenuation and phase constants, the intrinsic wave impedance and the wavelength in the tissue.
4. Power absorption in tissues was summarised and specific absorption rate (SAR) was defined.
5. Relationships between electric and magnetic fields across a boundary between two different tissues were outlined.

1.2 Dosimetry in Electromagnetic Hyperthermia

1.2.1 Phantom Materials

Phantom materials which simulate the dielectric properties of tissues play an important role in the develop-

Table 1.4. Gel-type muscle phantoms ($f \leq 100$ MHz)

Balzano et al. (1979)

f (MHz)	ε'	$\sigma(Sm^{-1})$	Composition (% by weight)			
			H_2O	NaCl	Al powder	Superstuff gelling agent
30	110	0.65	76.57	0.153	13.78	9.495

Bini et al. (1984)

f (MHz)	ε'	$\sigma(Sm^{-1})$	Composition per 100 ml					
			C_3H_5NO (g)	$C_7H_{10}N_2O_2$ (g)	$C_6H_{16}N_2$ (ml)	NaCl (g)	H_2O (ml)	$(NH_4)_2S_2O_8$ [a] (ml)
13.56	≈ 70	0.62	15	0.1	0.5	0.52	To 90	10
27.12	≈ 70	0.60	15	0.1	0.5	0.45	To 90	10
40.68	≈ 70	0.68	15	0.1	0.5	0.54	To 90	10

Chou et al. (1984)

f (MHz)	ε'	$\sigma(Sm^{-1})$	Composition (% by weight)			
			H_2O	NaCl	Al powder	TX-150 gelling agent
13.56	149	0.62	80.88	0.280	9.15	9.69
27.12	113	0.62	80.97	0.270	9.06	9.70
40.68	97.9	0.70	80.82	0.303	9.20	9.68
70	84.7	0.70	86.50	0.424	2.72	10.36
100	71.5	0.89	87.59	0.482	2.12	9.81

Kato and Ishida (1987)

f (MHz)	Composition (% by weight)		
	PVC powder	Agar	NaN_3
	$\dfrac{\text{PVC powder}}{\text{PVC powder} + H_2O}$	$\dfrac{\text{Agar}}{\text{Agar} + H_2O}$	$\dfrac{NaN_3}{NaN_3 + H_2O}$
1 – 40	$0 \leq P \leq 44.4$	4	$0 \leq N \leq 8 \times 10^{-3}$

$\varepsilon' = 79.25 - 1.009 \times P$
$\sigma = 4.144 \times 10^{-3} \times (146.6 \times N + 0.059) \times (85\text{-}P)^{1.235}$

[a] 1.3% by weight solution in H_2O

ment of electromagnetic applicators. By using such materials, the power deposition patterns of devices may be assessed in the absence of complex cooling mechanisms which are present in animal models or in patients. Quantitative or qualitative measurements may be made using invasive thermometry or some form of thermographic imaging. Recently, standard phantoms (Allen et al. 1988; Hand et al. 1989b) which enable realistic comparisons between devices to be made and quality assurance aspects of hyperther-

mia treatments to be developed, have been described. In addition to those phantom materials which simulate tissues with high or low water content (e.g. muscle and fat/bone), materials which simulate specific tissues (e.g. brain, lung) have been developed. The compositions and dielectric properties of several materials described in the literature are listed in Tables 1.4 – 1.8. Details of the density and specific heat for these materials and the temperature at which the dielectric properties were determined (usually around

Table 1.5. Gel-type muscle phantoms (f ≥ 100 MHz)

Andreuccetti et al. (1988)

f (MHz)	ε'	$\sigma(\mathrm{Sm}^{-1})$	Composition per 100 ml					
			C_3H_5NO (g)	$C_7H_{10}N_2O_2$ (g)	$C_6H_{16}N_2$ (ml)	NaCl (g)	H_2O (ml)	$(Na_4)_2S_2O_8$[a] (ml)
750	52	1.51	27	0.18	0.9	1.63	To 90	10
1050	49.5	1.66	30	0.20	0.2	1.75	To 90	10
2450	48	2.32	32	0.21	1.07	0.88	To 90	10

Chou et al. (1984)

f (MHz)	ε'	$\sigma(\mathrm{Sm}^{-1})$	Composition (% by weight)			
			H_2O	NaCl	Polyethylene powder	TX-150 gelling agent
200	56.7	1.06	74.92	0.894	15.79	8.39
300	54.8	1.17	75.15	0.996	15.44	8.42
433	53.5	1.21	75.15	0.996	15.44	8.42
750	52.5	1.26	75.15	0.996	15.44	8.42
915	51.1	1.27	75.15	0.996	15.44	8.42
2450	47.4	2.17	75.48	1.051	15.01	8.46

Hartsgrove et al. (1987)

f (MHz)	ε'	$\sigma(\mathrm{Sm}^{-1})$	Composition (% by weight)				
			H_2O	NaCl	Sugar	HEC[b]	Dowicil 75[c]
100	70.5	0.68	52.4	1.4	45.0	1.0	0.1
400	62.5	0.90	52.4	1.4	45.0	1.0	0.1
900	54.7	1.38	52.4	1.4	45.0	1.0	0.1

[a] 1.3% by weight solution in H_2O
[b] Hydroxyethylcellulose
[c] Bactericide

20 °C) as well as the methods for producing them can be found in most of the papers cited. Most phantom materials are solids or gels although a few liquid materials have been developed. The latter are compatible with using a mechanically scanned probe to determine field and hence power distributions.

A common method of performing SAR measurements is to determine the changes in temperature at known points in a solid or gel phantom following a brief period of heating at high power. With this procedure, effects of thermal conduction within the phantom are minimised and SAR may be inferred from

$$SAR = \frac{dT}{dt} c \, W kg^{-1} \qquad (1.55)$$

Table 1.6. Liquid muscle phantoms (f ≥ 100 MHz)

Gajda et al. (1979)

f (MHz)	ε'	$\sigma(\mathrm{Sm}^{-1})$	Composition (% by weight)	
			Glycerol	H_2O
2450	48	1.9	30	70

Hand et al. (1989a)

f (MHz)	ε'	$\sigma(\mathrm{Sm}^{-1})$	Composition (% by weight)		
			H_2O	NaCl	C_2H_5OH
450	56.0	1.2	63	1.5	35.5

Table 1.7. Fat-bone phantoms

Castable solid phantom: Guy (1971a)

f (MHz)	ε'	$\sigma(\mathrm{Sm^{-1}})$	Composition (% by weight)			
			Laminac polyester resin	Catalyst	Acetylene black	Al powder
915	5.61	0.067	85.20	0.375	0.24	14.5
2450	4.51	0.172				

Polyacrylamide phantom: Bini et al. (1984)

f (MHz)	ε'	$\sigma(\mathrm{Sm^{-1}})$	Composition per 100 ml					
			C_3H_5NO (g)	$C_7H_{10}N_2O_2$ (g)	$C_6H_{16}N_2O_2$ (ml)	NaCl (g)	$C_2H_6O_2$ (ml)	$(NH_4)_2S_2O_8$[a] (ml)
27.12	21.8	0.011	40	0.267	1.33	0.058	To 90	10

Dough phantom: Lagendijk and Nilsson (1985)

f (MHz)	ε'	$\sigma(\mathrm{Sm^{-1}})$	Composition (% by weight)		
			Flour	Oil	NaCl solution
451	7.3	0.038	66.7	30	3.3

Bone (castable material): Hartsgrove et al. (1987)

f (MHz)	ε'	$\sigma(\mathrm{Sm^{-1}})$	Composition (% by weight)		
			Epoxy	Hardener	KCl
100	13.6	0.08	35.0	35.0	28.0
400	9.3	0.110	35.0	35.0	28.0
900	7.4	0.16	35.0	35.0	28.0

Bone (liquid): Hartsgrove et al. (1987)

f (MHz)	ε'	$\sigma(\mathrm{Sm^{-1}})$	Composition (% by weight)				
			Tween	n-Amyl alcohol	Paraffin oil	H_2O	NaCl
100	10.8	0.035	57.0	28.5	9.5	4.5	0.5
400	9.1	0.066	57.0	28.5	9.5	4.5	0.5
900	7.2	0.12	57.0	28.5	9.5	4.5	0.5

Fat (liquid): Gajda et al. (1979)

f (MHz)	ε'	$\sigma(\mathrm{Sm^{-1}})$	Composition
2450	5.9	0.11	$CH_3(CH_2)_4Br$

[a] 0.65% by weight in ethanediol

where c $(\mathrm{J\,kg^{-1}\,°C^{-1}})$ is the specific heat of the phantom material. Although computer-controlled systems may be designed to make measurements of SAR distributions (e.g. Wong et al. 1985), the time required to complete such a measurement for an applicator of medium size or larger can be considerable since temperatures should be allowed to return to steady state following each power pulse to avoid a continuous increase in the temperature of the phantom throughout the experiment.

Table 1.8. Brain phantoms

Gel-type phantom: Hartsgrove et al. (1987)

f (MHz)	ε'	$\sigma(Sm^{-1})$	Composition (% by weight)				
			H$_2$O	NaCl	Sugar	HEC[a]	Dowicil 75[b]
100	63.0	0.47	40.4	2.5	56.0	1.0	0.1
400	50.3	0.75	40.4	2.5	56.0	1.0	0.1
900	41.2	0.122	40.4	2.5	56.0	1.0	0.1

[a] Hydroxyethylcellulose
[b] Bactericide

If the phantom is constructed so that it can be opened quickly to reveal a particular cross-section, a quantitative measurement of the two-dimensional distribution of SAR in this surface may be made using an infra-red camera (Guy 1971 a; Cetas 1982). Alternatively, liquid crystal sheets provide rapid and low cost qualitative assessment of distributions (Jones and Carnochan 1986). At microwave frequencies little error is introduced by covering the split surfaces in the phantom with thin plastic material (Schaubert 1984), but at lower frequencies (e.g. 27 MHz) the SAR distribution may be grossly distorted by the split unless precautions are taken to achieve electrical continuity across the split surfaces. Lehmann et al. (1983) have described the use of wetted silk screens for this purpose.

Another technique is to use an electric field probe within a liquid phantom. A simple probe consisting of a section of semi-rigid coaxial cable with a physically and electrically short length of central conductor exposed may be used for this purpose (Gajda et al. 1979). Such a probe will provide information on the field component parallel to the exposed section of central conductor. Information concerning the other components may be obtained by changing the direction of the section of exposed central conductor. If a network analyser is used in the measurement, phase distributions may also be determined. A drawback of the simple coaxial probe is that the presence of the coaxial feed may perturb the field distribution being measured. Also, artefacts may arise in the presence of boundaries between media with differing dielectric properties due to a change in the impedance of the probe caused by the change in dielectric environment. Coating the probe with a material for which ε' is less than that of the media reduces the problem (King and Smith 1981). A better method is to use a non-perturbing probe consisting of either a single or three mutually perpendicular dipoles, the latter arrangement allowing three orthogonal field components to be measured (Bassen and Smith 1983; Batchman and Gimpelson 1983).

1.2.2 Methods of Calculating SAR

In the context of electromagnetic hyperthermia, theoretical calculations of SAR distributions are useful in the design and assessment of applicators and in the planning and evaluation of clinical treatments. In the first of these applications it is often sufficient to consider analytical methods applied to simple and idealized models, e.g. homogeneous or layered half-spaces, circular cylinders or ellipses with dielectric properties of tissues, to gain insight into the behaviour of a particular device. On the other hand, studies relating to clinical treatments must take account of dielectric heterogeneities and the complex geometry of the body as well as realistic modelling of the electric and/or magnetic fields associated with the applicator used. These studies are likely to require the use of numerical methods. It is stressed that knowledge of the SAR distribution is only one step in planning and evaluation of clinical treatments. This information is required as input to a model which must take account of the thermal properties and transport mechanisms within tissues, particularly due to blood flow, to predict the resulting temperature distribution within the treatment volume.

In the following sections brief outlines of some of the methods used for predicting electric fields and SAR together with examples of their use in hyperthermia-related problems are given. For further details the reader is referred to Bardati (1986), Durney (1980, 1987), Paulsen (1989) and Spiegel (1984).

1.2.2.1 Analytical Models

A simple model which is a useful representation of local heating when the wavelength is short compared to the curvature of the tissues is that of an applicator in contact with a lossy homogeneous or layered half space. Examples of such models include a rectangular aperture (Guy 1971 b) and a circular aperture (Fray et al. 1982) in which the electric field within the tissue is evaluated using a Fourier transform technique (Harrington 1961). For example, in the bilayered model considered by Guy (1971 b) the electric fields in the fat and muscle tissue were found from:

$$\mathbf{E}_{f,m}(x,y,z) = \frac{1}{(2\pi)^2} \int\limits_{-\infty}^{\infty} \int\limits_{-\infty}^{\infty} \Phi_{f,m}(u,v,z)$$

$$\times \exp\left[j(ux+vy)\right] du\, dv \qquad (1.56)$$

where $\Phi_{f,m}$ are the Fourier transforms of the electric fields at the fat and muscle boundaries derived from the boundary conditions at the surface and at the fat/muscle interface in terms of $\Phi_a(u,v)$, the Fourier transform of the aperture field $\mathbf{E}_f(x,y,0)$, which is given by

$$\Phi_a(u,v) = \int\limits_{-\infty}^{\infty} \int\limits_{-\infty}^{\infty} \mathbf{E}_f(x,y,0) \exp\left[-j(ux+vy)\right] dx\, dy$$
$$(1.57)$$

An alternative method of determining the field distribution from an aperture source is to use the field equivalence theorem (Jordan and Balmain 1968) to replace the electric and magnetic fields $\mathbf{E}_a, \mathbf{H}_a$ in the aperture plane by equivalent magnetic and electric current surface densities $\mathbf{J}_{sm}, \mathbf{J}_{se}$ such that:

$$\mathbf{J}_{sm} = -\mathbf{n}\times\mathbf{E}_a \quad \text{and} \quad \mathbf{J}_{se} = -\mathbf{n}\times\mathbf{H}_a \qquad (1.58)$$

where $\mathbf{n}$ is a unit vector normal to the aperture plane and directed into the lossy medium. $\mathbf{J}_{sm}$ and $\mathbf{J}_{se}$ are in turn represented by an array of point sources or infinitesimal dipoles. For each dipole the magnetic and electric vector potentials $\mathbf{A}$ and $\mathbf{F}$ at a point of interest at distance r are:

$$\mathbf{A} = \frac{\mu_0 \mathbf{J}_{se} \exp\left[-\gamma r\right]}{4\pi r} \quad \text{and} \quad \mathbf{F} = \frac{\varepsilon \mathbf{J}_{sm} \exp\left[-\gamma r\right]}{4\pi r} \qquad (1.59)$$

with $\gamma^2 = -\omega^2 \mu_0 \varepsilon$. The electric and magnetic fields due to a point source are:

$$\mathbf{E} = -j\omega\mathbf{A} - \frac{j}{\omega\mu_0\varepsilon}\nabla(\nabla\cdot\mathbf{A}) - \frac{1}{\varepsilon}\nabla\times\mathbf{F} \qquad (1.60)$$

$$\mathbf{H} = -j\omega\mathbf{F} - \frac{j}{\omega\mu_0\varepsilon}\nabla(\nabla\cdot\mathbf{F}) - \frac{1}{\mu_0}\nabla\times\mathbf{A} \qquad (1.61)$$

and the total electric field at any point is found by vector addition of the contributions from each elemental source in the aperture plane. This method has been used by a number of authors to investigate

penetration due to aperture sources in homogeneous media (e.g. Turner and Kumar 1982; Lagendijk 1983; Nilsson 1984; Hand et al. 1986) and in layered media (e.g. Johnson et al. 1987).

Bach Andersen (1986) and Lumori (1988) have investigated the use of Gaussian beams to calculate fields in homogeneous and layered (plane and curved interfaces) media from aperture sources. Although the model is a high frequency approximation, results indicate that fields from applicators with dimensions comparable with the wavelength in lossy media such as muscle or lung phantoms can be simulated. An attractive feature of the Gaussian beam method is the capability of computing three-dimensional fields in a short CPU time.

Cylindrical models in which the electric or magnetic field is specified at the surface of the cylinder have been used to model small aperture sources (e.g. Ho et al. 1971), arrays of aperture sources (Wait and Lumori 1986) and generic applicators (e.g. Brezovich et al. 1982; Morita and Bach Andersen 1982). Aperture field distributions at the surface of the cylinder may be expressed as a two-dimensional Fourier series and the fields within the cylinder as a summation of three-dimensional cylindrical waves (Wait 1959).

1.2.2.2 Numerical Models

Method of Moments. The method of moments consists in transforming an integral equation which describes the field within the body and is derived from Maxwell's equations with appropriate boundary conditions into a matrix equation and obtaining a solution by matrix inversion (Harrington 1968). In this method the body is assumed to comprise a collection of polygons, the simplest case being to use cubic blocks and to assume that the field within any block is constant (a pulse basis function) (Livesay and Chen 1974; Chen and Guru 1977). The basis functions are substituted into the original integral equation and the resulting integral equation is then transformed into a matrix equation by multiplication by a set of weighting functions. The simplest weighting functions are delta functions, which enable the field to be calculated at the centre of each block. In Galerkin's method the weighting functions are taken to be equal to the basis functions. Massoudi et al. (1984) discussed the limitations of pulse functions in satisfying boundary conditions and suggested other techniques which might provide better modelling of complex body shapes and give a smoother representation of the field, albeit at the expense of increased complexity. In a later study Tsai et al. (1986) investigated the use of arbitrarily shaped polyhedra volume elements and three-dimensional linear basis functions for bodies with simple geometry. Although in principle a solution can be specified for any type of body and external field distribution, it may not be possible in practice to obtain the solution because of the large order matrix involved. For a model consisting of N cells, requirements for computer storage and computation time are proportional to $(3N)^2$ and $(3N)^3$, respectively. Borup and Gandhi (1984, 1985) discuss a fast Fourier transform moment method for which storage and computation time are proportional to N and $N \log_2(N)$ for a uniform mesh. Other iterative methods have been described by Kastner and Mittra (1983), van den Berg (1984), and Sultran and Mittra (1985). In general these are aimed at reducing storage requirements, sometimes at the expense of run time, which is very important in three-dimensional modelling.

The moment method has been applied to hyperthermia-related problems by several authors. For example, Iskander et al. (1982) used the method to calculate two-dimensional distributions of absorbed power density in the human thorax due to an annular array applicator. Hill et al. (1983) used the moment method with Galerkin's method in a two-dimensional calculation of absorbed power density in the thorax due to a concentric coil applicator. van den Berg et al. (1983) used their iterative method to calculate the fields within a cross-section of the pelvic region due to two opposed 27-MHz ridged waveguides driven in phase. A three-dimensional 180 cell block model of man was used by Hagmann and Levin (1986) to calculate SAR within the body due to exposure of the abdominal region to a longitudinally polarised electric field at frequencies from 10 to 60 MHz.

Finite Difference Time Domain (FDTD) Method. In this method a grid is superimposed on the body and surrounding region and components of **E** and **H** are found at locations within each unit cell by following the propagation, scattering and absorption of the fields from the source through the grid in a time-stepping manner by means of a finite difference solution to the differential form of Maxwell's equations (Yee 1966; Taflove and Brodwin 1975a; Lau and Sheppard 1986; Sullivan et al. 1987). At the boundaries of the finite grid the outgoing wave must be prevented from being reflected back into space occupied by the grid. One method of achieving this is to impose an absorption condition at the edge of the grid. Since the values of the E or H components at a grid point at any time step depend only on the previous value of that component and the previous values of the neighbouring H or

E components, respectively, a solution to a large matrix equation is not required. The computer storage and time requirements are proportional to N, the number of cells. This is a major advantage over the integral equation techniques even though more cells may be required to implement the FDTD method than the moment methods. With present computing power, calculations are restricted to frequencies between about 50 and 700 MHz (Sullivan et al. 1988). At low frequencies more time steps and a larger total space are required whilst higher frequencies require a smaller grid size and therefore a larger number of cells.

An early application of this technique to a biomedical problem was the computation of electromagnetic fields and temperature distributions in the human eye subjected to microwave radiation (Taflove and Brodwin 1975 b). More recently Lau (1986) and Lau et al. (1986) have used the method in three-dimensional modelling of a 434-MHz microwave lens applicator for local hyperthermia and Wang and Gandhi (1989) simulated annular phased arrays of aperture or dipole antennas driven at 100 MHz placed around an anatomically based model of the human torso.

Finite Difference Frequency Domain Methods. Finite difference calculations in the frequency domain have also been used in hyperthermia-related problems. Doss (1982) used a finite difference formulation of Laplace's equation to find potential, electric field and SAR distributions in simple models representing RF capacitive heating. Armitage et al. (1983) represented anatomically real cross-sections of the body by an equivalent admittance network. The potential at each node point was determined by solving the admittance matrix equation by successive overrelaxation. The electric field and SAR were then found using a finite difference technique. The method was applied to problems involving RF capacitive electrodes and a concentric coil applicator. Gandhi et al. (1984) described a similar method in which an equivalent impedance network was used. The impedance of each element is determined by the local dielectric properties and anisotropy in σ and ε' within the tissues can be taken into account. These authors used their technique to calculate SAR within a realistic body cross-section due to various axially directed RF magnetic fields.

The quasi-static assumptions made in all three studies require that the wavelength be considerably larger than the dimensions of the tissues or body cross-section and therefore the methods may be used for frequencies up to approximately 40 MHz.

Finite Element Method. In the finite element method (Strohbehn et al. 1986; Paulsen 1989), the body and surrounding space are covered by a series of cells referred to as elements which may be either domain or boundary types and which are interconnected at points known as nodes. The equations $\nabla \times \mathbf{E} = j\omega\mu\mathbf{H}$ and $\nabla \times \mathbf{H} = -j\omega e\mathbf{E}$ are combined to give:

$$\nabla \times \left(\frac{1}{j\omega\mu} \nabla \times \mathbf{E} \right) + j\omega\varepsilon\mathbf{E} = 0 \qquad (1.62)$$

and a similar equation in $\mathbf{H}$. These equations are then multiplied by weighting functions and integrated. The unknown $\mathbf{E}$ is written in terms of the basis functions, boundary conditions are imposed and $\mathbf{E}$ is determined at each node or vertex of the finite element grid by solving a matrix equation. Values elsewhere within the cell are determined in terms of the basis function. An advantage of the finite element technique is that tissue heterogeneities, irregular boundaries and variable spatial resolution may be readily accounted for. Disadvantages of the technique are that a large matrix equation must be solved and that determination of suitable boundary conditions at the outer boundary of the finite element grid can be difficult.

The finite element technique has been used to investigate the electric field and SAR distributions within cross-sections derived from CT scans of patients for two hyperthermia devices − the concentric coil and the annular array applicator (Paulsen et al. 1984b, 1985; Strohbehn et al. 1986). Calculations in three dimensions using the finite element method and the boundary integral method (Paulsen et al. 1988a) have been made for the concentric coil, the annular array applicator and RF capacitive plate electrodes (Paulsen et al. 1988b).

1.2.3 Summary

1. The use of phantom materials as an aid to applicator development and quality assurance was outlined.
2. Recipes for several types of phantom material (gel, solid, liquid) which simulate muscle, fat, bone and some specific tissues were given.
3. A brief review of analytical and numerical techniques used for calculating SAR distributions was presented.

1.3 Electromagnetic Techniques for Hyperthermia

1.3.1 Overview of Techniques

The variety of tumour sizes, shapes and locations presented by patients implies that several types of hyperthermia system, each with its own characteristics, are likely to be required. The selection of a particular system or applicator is usually determined by quality assurance requirements, e.g. the minimum requirement is that the useful field of the applicator must cover the tumour volume. A detailed discussion of the topic of quality assurance is given in the chapter by Dr. Shrivastava in this volume.

From physical and engineering points of view, electromagnetic applicators are antennas which couple power into lossy media (the tissues). Insight into the behaviour of applicators can be gained by considering their dimensions and those of the tissues to be heated in terms of the wavelength. For example, if an applicator has dimensions which are comparable with or greater than half the wavelength, it will radiate electromagnetic waves which will be attenuated and reflected or transmitted as they propagate through the inhomogeneous tissues. On the other hand, an applicator whose dimensions are small compared with the wavelength is a poor radiator and the structure of the fields near to the applicator must be given further consideration. Unlike plane waves, the electric and magnetic fields in this region are not orthogonal and the ratio E/H is different from that of free space and may vary from point to point. Close to the applicator the fields are relatively intense but they decrease more rapidly with increasing distance from the applicator than the exponential decrease associated with plane waves. Thus excessive heating of superficial tissues may result and the observed penetration depth will be smaller than that calculated for plane waves.

Morita and Bach Andersen (1982) made an analytical study of near-field coupling of electromagnetic sources to circular cylinders with dielectric properties similar to those of muscle. Elementary electric and magnetic dipoles at frequencies below 300 MHz were spaced from the longitudinal (z) axis of the cylinder at distances which were much smaller than the free-space wavelength and orientated with each of three polarisations (x, y, z) relative to the axis. The wavelengths and penetration depths at these frequencies were large compared with the radius of the cylinder. Using this long wavelength approximation, they showed that the axially (z-) aligned electric dipole source achieved the greatest penetration, a finding of relevance in the design of applicators for regional hyperthermia, whilst the x- and y-oriented electric dipoles produced distributions of absorbed power which were highly dependent upon the separation between source and cylinder. Such dependence was less critical in the case of the y-directed magnetic dipole. The x-directed magnetic dipole produced an absorbed power distribution with zero along that axis. These authors drew attention to the fact that near-field fall off from finite sized aperture sources, which can be considered as distributions of these elementary sources, is less rapid than for elementary sources and is dependent upon aperture size.

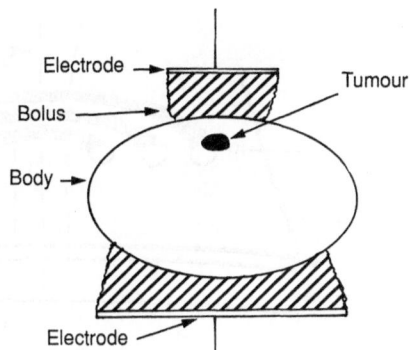

Fig. 1.2. Capacitive or E-field applicator

1.3.2 Techniques for Local (Superficial) Hyperthermia

At frequencies lower than about 40 MHz the wavelength is considerably larger than both the applicator and the tissues to be heated. In these cases the problem may be considered as being "quasi-stationary" in that the spatial variations of the electric and magnetic fields can be taken to be the same as those of static (dc) fields. Examples include capacitive electrodes and inductive coils, techniques which have been applied for many years in physiotherapy. At these frequencies, energy deposition in the tissue is due predominantly to resistive heating by conduction currents ($\tan \delta > 1$). It is emphasised that the fall off in field strength with increasing distance from these electrically small applicators is much more rapid than is the case for plane waves at the same frequencies. Indeed the penetration from these devices is determined more by geometry than by operating frequency. Although applicators may be designed to produce a distribution of energy deposition with some general characteristics in simple phantom materials, clinical performance in the electrically inhomogeneous human body is difficult to predict.

Capacitive or E-field applicators usually consist of two electrodes placed on the body such that the tumour lies between them, as shown in Fig. 1.2. The construction of the electrodes usually includes an integral bolus which enables them to conform to some degree to body contours. Their size and shape may be chosen independently of frequency although small electrodes are only useful for treating superficial tumours.

Inductive or H-field applicators can be considered as magnetic dipoles. For the traditional techniques employed in physiotherapy the dipoles are orientated either perpendicular to the body surface (as in the case of a "pancake coil") or coaxially with the body [as for a current loop around the body (Fig. 1.3)]. Recently devices in which the magnetic dipole is parallel to the body surface have been developed. This orientation of the magnetic dipole avoids the axial null in energy deposition in the tissue which is a characteristic of pancake coils or current loops. The size and shape of inductive applicators are not closely related to the operating frequency. Unlike capacitive applicators, magnetic applicators do not require contact with the body, which is sometimes an advantage in clinical use.

When the dimensions of an applicator are comparable with the half-wavelength, energy may be radiated from it. At frequencies greater than about 400 MHz, waveguide apertures or horns (Fig. 1.4) were usually chosen to transmit energy to the tissues. Recently, other wave-guiding structures such as microstrip have been used to produce smaller applicators with a considerable reduction in weight. Some of these new designs are conformable to body contours whilst others have enabled a compact device to be operated at frequencies of 200 MHz and lower. Most radiating applicators require the use of a bolus to space the tissue from the relatively rapidly decaying near-field. The presence of the bolus also improves the impedance match to the tissue and reduces the level of stray radiation compared with the case of an air gap between applicator and tissue as well as providing a means of cooling the skin and other superficial tissue.

Experience has shown that use of a single applicator, be it capacitive, inductive or radiative, leads to a number of problems when treating superficial tumours. For example, the distribution of absorbed power cannot be changed in a predictable manner during treatment in an attempt to optimise the observed temperature distribution. The operator can merely adjust the power delivered to the applicator and change the posi-

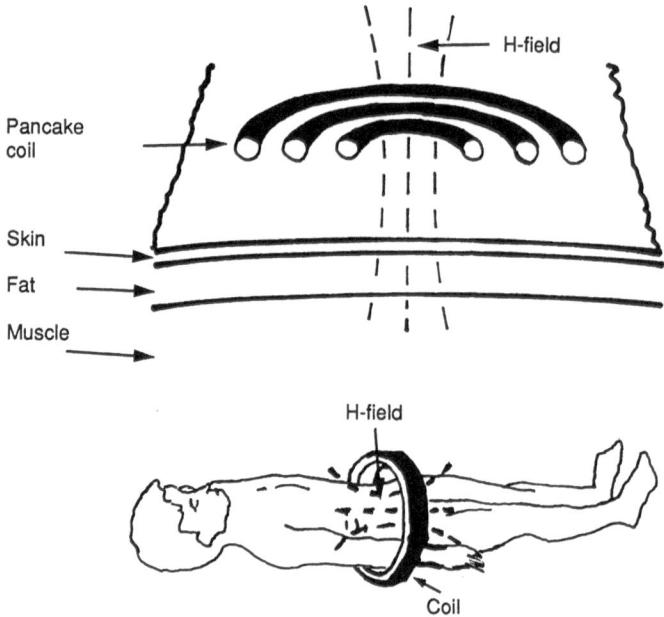

Fig. 1.3. Examples of inductive or H-field applicators. The *upper diagram* shows a "pancake coil" which produces an H-field which is essentially normal to the superficial tissue interfaces. The *lower diagram* shows a current loop concentric with the body which produces an H-field which is predominantly parallel with the patient's longitudinal axis

tion of the applicator relative to the patient. The use of an array of applicators in which the power delivered to individual devices could be controlled should improve this situation. In the treatment of some tumours in the neck a modest increase in penetration over that possible with a single applicator would be very useful. This may be possible using an array of applicators with relative phase and amplitude control. Many patients present with recurrences on the chest wall with tumour involvement over an area which is too large to treat using a single applicator. The depth of tumour involvement is often quite superficial and many of these patients could be offered better hyperthermal treatment if conformable arrays of electromagnetic applicators were available. Investigations into microwave arrays have used waveguide apertures or horns, microstrip-based and other small applicators. The choice of frequency and phase and amplitude driving conditions is an active area of research.

1.3.3 Techniques for Regional (Deep) Hyperthermia

Non-invasive electromagnetic techniques are unable to deposit energy selectively in deep-seated tumours. This results from the need to use a relatively low frequency (100 MHz or lower) to avoid unacceptably high attenuation when traversing superficial tissues. The wavelength in high water content tissues at these

frequencies is 27 cm or longer and so the best possible "focus" would have dimensions of at least 14 cm. The concept underlying the use of an electromagnetic deep heating device is to deposit energy within the general region of interest and to rely upon physiological factors such as low rate of blood perfusion in the tumour relative to normal tissue to produce higher temperatures in the tumour compared with those in surrounding normal tissue. Theoretical studies of several devices and experience gained using the same devices in the clinic have subsequently shown that the original concept of regional hyperthermia was optimistic. Induction of therapeutic temperatures in deep tumours non-invasively by electromagnetic fields remains a major problem.

The device which have been considered for regional hyperthermia can by broadly classified in terms of the direction of the E-field produced within the tissue in

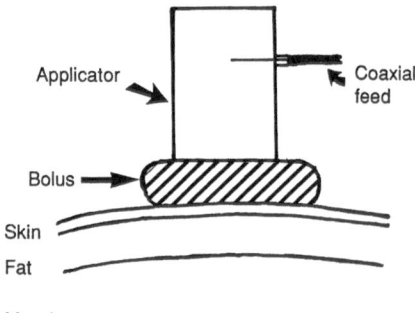

Fig. 1.4. A waveguide applicator with bolus

relation to the body surface and tissue boundaries. In one class, devices produce an electric field which is predominantly perpendicular to the skin and superficial tissues (Fig. 1.5). As described in Sect. 1.3.2, this is a characteristic of capacitive (electric) applicators. In Japan, interest in capacitive techniques has led to the development of machines which employ large electrodes (typically 20 or 25 cm in diameter) and operate at 13.56 or 8 MHz. The concept of this approach is that in clinical use the electrodes are separated by a distance comparable with, or smaller than, their diameter and so might be expected to produce significant power densities throughout the region between them. The problems of excessive power deposition in the low permittivity adipose tissue and the inability to modify absorbed power distributions from a two-electrode system referred to in Sect. 1.3.2 also limit the effectiveness of the capacitive technique when applied to regional hyperthermia.

In another class of devices, the E-field produced within the tissue is parallel to skin and superficial tissue boundaries but not aligned with the patient's longitudinal axis. One example is the current loop placed concentrically around the patient (Fig. 1.6). As with all loop devices there is zero power deposition at the centre of the loop, which in this case is on, or near to, the patient's central axis. Subsequent theoretical studies and clinical experience have shown that this type of device is not effective for treating tumours located deep in the body. Other current loop orientations are possible, for example coaxial coils placed above and below the patient. This approach does produce finite power deposition at the patient's central axis but there is a null on the common axis of the coils.

In a third class of devices the E-field is parallel to the skin and superficial tissues and is aligned with the patient's longitudinal axis (Fig. 1.7). Several devices are designed to produce such a field around the circumference of the patient. With this configuration it is possible to achieve constructive interference leading

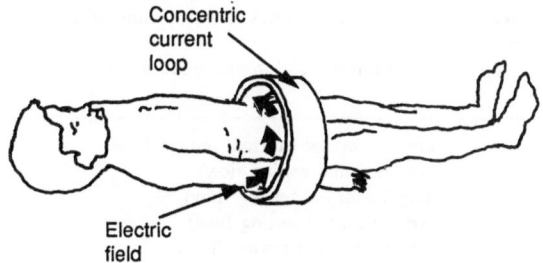

Fig. 1.6. Electric field produced by a concentric coil applicator

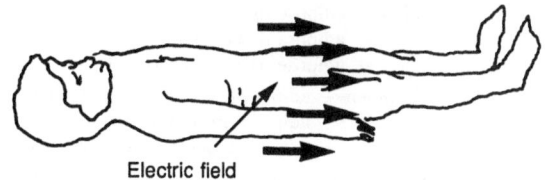

Fig. 1.7. Longitudinally directed circumferential electric field

to significant power deposition at, or near to, the patient's central axis. Devices in this class include an annular array of aperture sources, a TEM device with a circumferential aperture, a concentric segmented cylindrical array and self-resonant helices. With some of these devices it is possible to modify the absorbed power distribution by adjusting the relative phases and amplitudes of the fields from individual sources.

1.3.4 Power Requirements for Hyperthermia Systems

The temperature distribution produced within tissues is determined by the balance between absorbed power and rate of energy loss by thermal conduction and the effects of blood perfusion. Using the bioheat equation:

$$\varrho c \frac{\partial T}{\partial t} = \nabla k \nabla T + w c_b (T_a - T) + M + P \qquad (1.63)$$

where ϱ, k, c and T are the density (kg m^{-3}), thermal conductivity (W m^{-1} °C^{-1}), specific heat (J kg^{-1} °C^{-1}) and local temperature (°C) of the tissue, respectively. c_b (J kg^{-1} °C^{-1}) is the specific heat of the blood, T_a is the arterial blood temperature, t is time (s), w is the local tissue perfusion (kg m^{-3} s^{-1}), M is the local metabolic rate (W m^{-3}) and P is the local power deposition rate (W m^{-3}) ($P = \varrho \cdot$ SAR). For our purposes we shall neglect M,

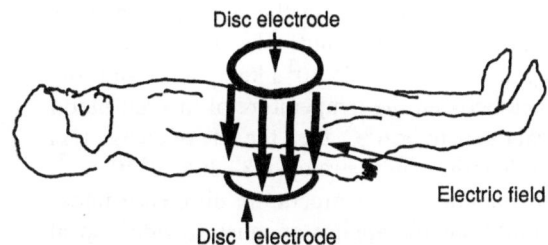

Fig. 1.5. Capacitive applicator producing an electric field predominantly normal to superficial tissue boundaries

Table 1.9. Measured blood perfusion rates of various normal human tissues and organs (compilation from Sekins and Emery 1982)

Tissue	Anatomical location and qualifications	Perfusion $(kg\,m^{-3}\,s^{-1})$
Skin	Foot – dorsal (normal resting flow)	2.38 ± 0.43
	Calf (normal resting flow)	1.77 ± 0.22
	Thigh (normal resting flow)	1.63 ± 0.43
	Arms (normal resting flow)	1.40
	Hands (normal resting flow)	3.34
	Abdomen (normal resting flow)	1.44
	Thorax (normal resting flow)	1.08
	Head (normal resting flow)	7.15
	Face (normal resting flow)	11.72
Subcutaneous fat	Thigh – adipose tissue 11 mm thick	0.93
	– adipose tissue 20 mm thick	0.33
	– adipose tissue 43 mm thick	0.15
	Abdomen – adipose tissue 10–29 mm thick	0.51 ± 0.35
	– adipose tissue 30–39 mm thick	0.36 ± 0.20
	– adipose tissue > 40 mm thick	0.31 ± 0.12
Muscle	Anterior calf (resting flow)	0.46 ± 0.12
	Anterior thigh (resting flow)	0.43 ± 0.17
	Forearm (resting flow)	0.53 ± 0.23
Bone	Range of flow in humerus (marrow only)	$0.06 \rightarrow 0.11$
Organ	Brain	9.0
	Liver	9.62
	Heart	14.0
	Kidney	70.0

Table 1.10. Measured blood perfusion rates of various human tumours (compilation from Jain and Ward-Hartley 1984)

Tumour	Perfusion $(kg\,m^{-3}\,s^{-1})$
Liver carcinoma	2
Adenocarcinoma of breast	5 –6.7
Corpus uteri (various)	1.7–6.7
Lymphangiomas	1.7
Lymphomas	5.5 ± 4
Anaplastic carcinomas	1.7 ± 1.3
Differentiated malignomas	3.7 ± 2.5
Glioma	5.5–15.3
Brain (metastatic primary)	4.2–9.2
Brain (metastatic carcinomas)	< 8.3
Brain (various)	≈ 8.3
Primary tumours	5 –8.3

which is on the order of $1000\,W\,m^{-3}$ ($1\,W\,kg^{-1}$), in comparison with P, which is approximately two orders of magnitude greater.

At the beginning of a local treatment a rate of increase in temperature of approximately $1\,^\circ C$ per minute is desirable. Initially,

$$\varrho c \frac{\partial T}{\partial t} = P \qquad (1.64)$$

and so taking $\varrho = 10^3\,kg\,m^{-3}$, $c = 3.5 \times 10^3\,J\,kg^{-1}\,^\circ C^{-1}$ and a rate of increase in temperature of $0.017\,^\circ C\,s^{-1}$, we find that P must be approximately $6 \times 10^4\,W\,m^{-3}$ (SAR $= 60\,W\,kg^{-1}$). As the treatment proceeds, a decrease in the local net rate of energy deposition will occur due to blood perfusion and thermal conduction and so increased power is required to maintain a suitable rate of increase in temperature. Indeed, the power required to overcome losses due to perfusion is much greater than that which is needed to produce an initial $\partial T / \partial t$ of $1\,^\circ C$ per minute. In general we do not have a detailed knowledge of the blood perfusion and it can be seen from Tables 1.9 and 1.10, which list some blood perfusion rates for various human tissues and tumours reported in the literature, that the range of perfusions is large. If we consider the second term on the right-hand side of Eq. (1.63), $wc_b(T_a - T)$, and take w in the range $2 - 15\,kg\,m^{-3}\,s^{-1}$, $c_b = 4 \times 10^3\,J\,kg^{-1}\,^\circ C^{-1}$ and the difference between the temperature of arterial blood and local tissue to be $6\,^\circ C$, then the rate of energy loss due to this term is in the range $4.8 - 36 \times 10^4\,W\,m^{-3}$. Thus to maintain this temperature difference under these conditions the applicator must provide a local SAR in the range $48 - 360\,W\,kg^{-1}$, values toward the high end of this range being appropriate to some

Table 1.11. RF/microwave power requirements for a local hyperthermia system

Frequency (MHz)	Generator output (W)
2450	100
915	200
433	400
146	500
27	500

brain tumours with high perfusion. Lehmann et al. (1978) considered blood flow in human muscle heated by a 915-MHz direct contact applicator to be approximately 30 ml/100 g/min ($5 \text{ kg m}^{-3} \text{ s}^{-1}$) and found that the SAR required to raise temperature to the therapeutic range in this case was $120-170 \text{ W kg}^{-1}$. Bassen and Coakley (1981) suggested that applicators intended for therapy should be cable of producing SAR values of 235 W kg^{-1} in the muscle region of a muscle/fat phantom.

The total power requirement is dependent upon the volume of tissue treated (the effective field size of the applicator) and the efficiency of the applicator. In addition power losses occur in the bolus and in the transmission line from the generator to the applicator. Table 1.11 lists power requirements which are typical of specifications of commercial systems using one or two applicators for local hyperthermia.

In the case of systems which use an array of applicators a choice between coherent and incoherent operation must be made. For quasi-planar arrays, for example as might be used in the treatment of large

areas on the chest wall, coherence is probably not required and the output of the generator could be switched sequentially from one applicator to another at a rate which is fast compared with thermal processes occurring in the tissues. Alternatively, each applicator could be driven by its own signal and power source (Fig. 1.8). When constructive interference is desired in some part of the field, the applicators must be driven coherently. This may be achieved by dividing the power from a single generator between applicators. A means of adjusting the relative phase and power to each applicator should be included. Controlling the average power delivered to applicators by switching power from an applicator to a 50-ohm termination is not possible in a coherent system using power dividers. An alternative approach is to drive each applicator from individual power amplifiers fed from a common signal source (Fig. 1.9). Control of the average power delivered to the applicators must be achieved by varying the amplitudes and not by varying duty cycles. Care should be taken in setting the relative phases in a coherent system which includes tuners since adjusting a tuner could change the phase of the fields from the corresponding applicator. Often the individual applicators used in an array will be smaller than applicators designed to be used individually, in which case an appropriate power for each generator is about 100 W.

Since there is little frequency dependence of penetration depth associated with most applicators used for local hyperthermia (say with dimensions in the range 5 cm×5 cm up to 15 cm×15 cm), it is sufficient in

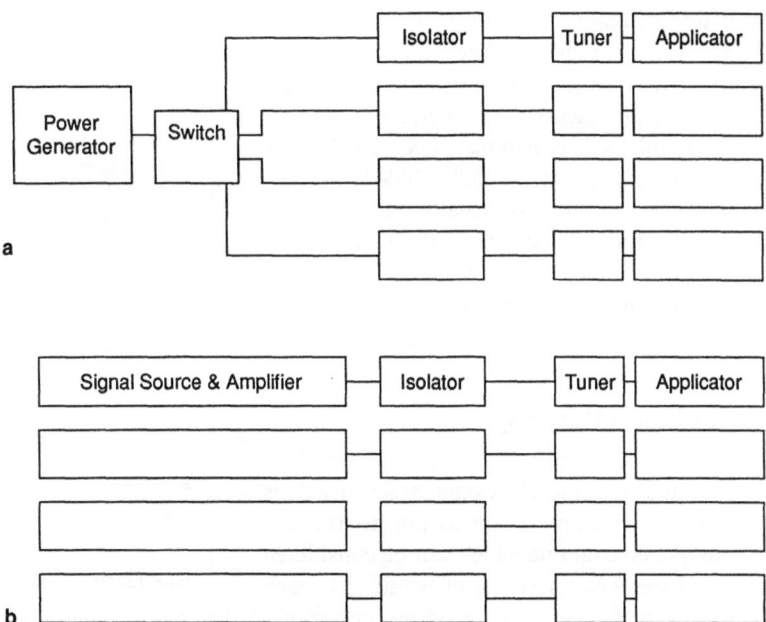

Fig. 1.8 a, b. Incoherent multichannel systems. **a** Incoherent system in which power is switched between channels. **b** Incoherent system using individual amplifiers

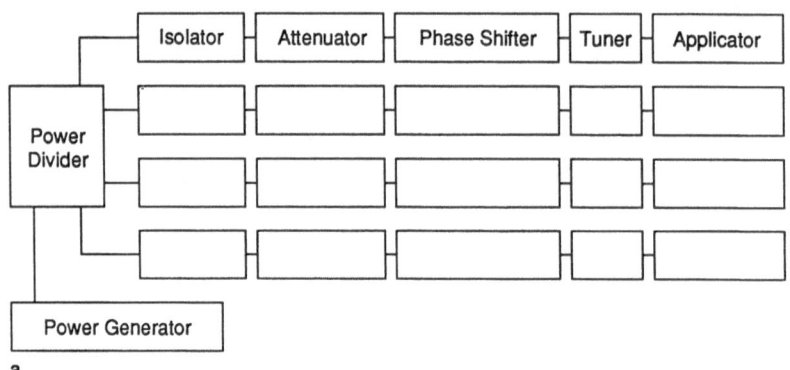

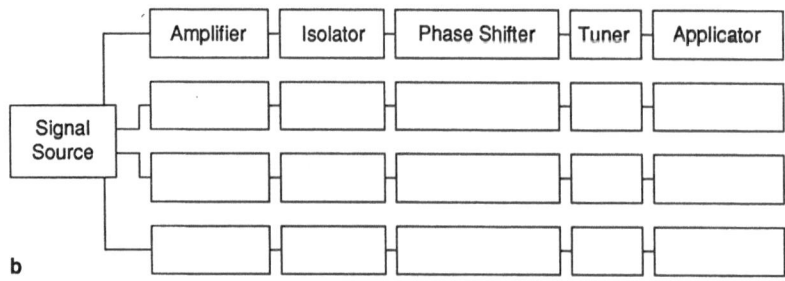

Fig. 1.9a, b. Coherent multichannel systems. **a** Coherent system using a single source and a power divider. **b** Coherent system using individual amplifiers and a common signal

clinical applications of local hyperthermia to use one or two generators, each providing power over a narrow band of frequencies centred at one of the frequencies listed in Table 1.11. In this way expensive variable frequency wide band power sources are avoided. However, wide band generators are useful in the physics/engineering laboratory for developing new devices.

In the case of regional hyperthermia devices, the requirements are different. For capacitive applicators a narrow band generator operating at 8, 13.56 or 27.12 MHz is usually used. Since a much larger volume of tissue is heated during regional hyperthermia, power requirements are increased and a generator capable of providing 1000–1500 W is usually used. A total power of 1500–2000 W from a variable frequency source is usually provided for radiative applicators. In the case of an array applicator this total may be provided by several individually controlled coherent sources.

1.3.5 Impedance Matching

In hyperthermia systems electromagnetic power is usually transmitted from the generator to the applicator along a coaxial line which can be considered as a series of elemental sections of length Δx, each represented by a T-network of a series impedance

$Z = R + j\omega L$ and a shunt admittance $Y = G + j\omega C$ as shown in Fig. 1.10. In this figure $R = R'(\Delta x/2)$,

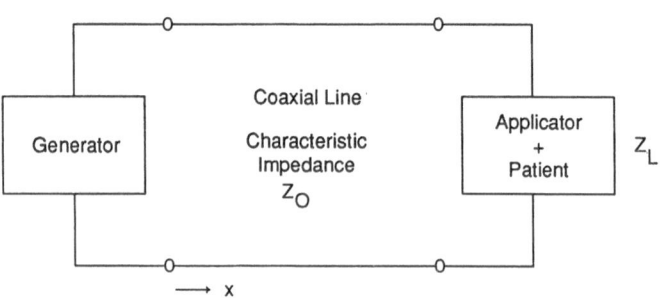

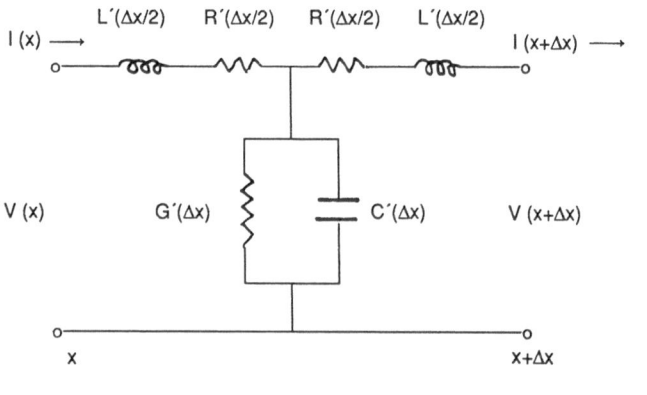

$R = R'(\Delta x/2)$ $L \doteq L'(\Delta x/2)$ $G = G'(\Delta x)$ $C = C'(\Delta x)$

Fig. 1.10. Transmission line and equivalent circuit

$G = G'(\Delta x)$, $L = L'(\Delta x/2)$, and $C = C'(\Delta x)$, where $R'(\Omega\,m^{-1})$, $G'(\Omega^{-1}\,m^{-1})$, $L'(H\,m^{-1})$ and $C'(F\,m^{-1})$ are the resistance, conductance, inductance and capacitance per unit length of the line, respectively. It is convenient to consider voltage and current waves along the line rather than electric and magnetic fields. The voltage $V(x)$ and current $I(x)$ along the line satisfy:

$$V(x) = V_1 \exp\,[-\gamma x] + V_2 \exp\,[\gamma x] \qquad (1.65)$$

and

$$I(x) = \left(\frac{Y}{Z}\right)^{1/2} (V_1 \exp\,[-\gamma x] - V_2 \exp\,[\gamma x]) \qquad (1.66)$$

where $\gamma = (ZY)^{1/2}$ is the complex propagation constant.

If there is a discontinuity in the line a reflection will occur $(V_2 \neq 0)$. The voltage reflection coefficient $\Gamma(x)$ is given by:

$$\Gamma(x) = \frac{V_2 \exp\,[\gamma x]}{V_1 \exp\,[-\gamma x]} \qquad (1.67)$$

or in terms of the characteristic impedance of the line $Z_0 = (Z/Y)^{1/2}$ and the impedance of the load (i.e. applicator plus patient) Z_L:

$$\Gamma = \frac{Z_L - Z_0}{Z_L + Z_0} \qquad (1.68)$$

The input impedance at any point along the line is:

$$Z_{in} = \frac{V(x)}{I(x)} = Z_0 \frac{(1+\Gamma)}{(1-\Gamma)}$$

$$= Z_0 \frac{Z_L \cosh\,(\gamma x) + Z_0 \sinh\,(\gamma x)}{Z_0 \cosh\,(\gamma x) + Z_L \sinh\,(\gamma x)} \qquad (1.69)$$

The impedance of the applicator is in general different from the characteristic impedance of the coaxial line (which is usually 50 ohm). In this case a reflection occurs at the applicator and a standing wave pattern is produced along the line. The ratio of maximum to minimum voltage along the line is known as the standing wave ratio (SWR) and is given by:

$$SWR = \frac{1+|\Gamma|}{1-|\Gamma|} \qquad (1.70)$$

Conversely:

$$|\Gamma| = \frac{SWR-1}{SWR+1} \qquad (1.71)$$

Significant reflected power is undesirable not only because the efficiency of the system decreases but also because it may prove detrimental to the generator. For optimal power transfer $\Gamma = 0$ and SWR = 1, i.e. Z_L must be matched to Z_0. At the lower RF frequencies used for hyperthermia an impedance matching network with lumped components may be employed (Fig. 1.11).

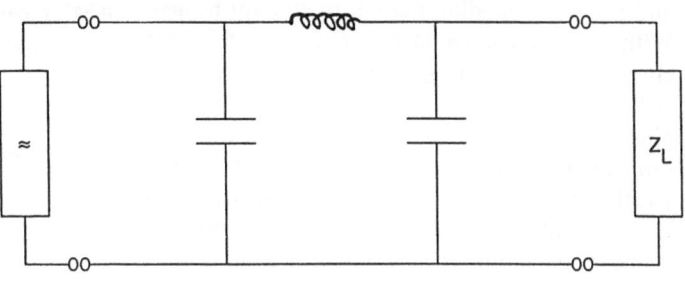

π - Section network

Stub tuner

Fig. 1.11. Impedance matching techniques

At the higher frequencies used for radiative applicators impedance matching may be achieved using a stub tuner. This is a shorted low loss transmission line whose length can be varied through the use of telescoping conductors. If we assume the line has no loss, then using Eq. (1.69) with $Z_L = 0$ we find that the input impedance of the shorted line is:

$$Z_{st} = j Z_0 \tan (\beta l) \qquad (1.72)$$

i.e. a pure reactance which is capacitive or inductive depending upon l, the length of the stub. With proper adjustment of l and its position d from the load (applicator) (Fig. 1.11) the line may be matched to the impedance produced by the parallel combination of the stub and length d of line terminated by the mismatched load Z_L consisting of applicator and patient. In practice the stub tuner should be placed as close to the applicator as possible to avoid high SWR and possible overheating in the mismatched section of line.

It is also advisable to include isolators in the system so that interaction between channels is reduced and generators are protected should significant power be reflected from the applicator/patient load (Fig. 1.8, 1.9).

1.3.6 Summary

1. Two types of hyperthermia treatment were considered. During local hyperthermia the tumour and some surrounding normal tissues are heated. With the non-invasive methods considered in this chapter, treatment is generally limited to superficial tumours within 3 cm of the skin. During regional hyperthermia energy is deposited throughout a large volume of tissue and differences in electrical properties and blood flow between tumour and normal tissues are relied upon to produce higher temperatures in the tumour than in normal tissue.
2. Three types of electromagnetic applicator were introduced. These were electric (capacitive plates), magnetic (induction coils and other current-carrying structures) and radiative (waveguide apertures or horns, microstrip radiators). The characteristics of all types of applicator are determined by the dimensions relative to the wavelength at the operating frequency. Detailed discussions of applicators are given in Sects. 1.4 and 1.5.
3. The power supplied to an applicator must be sufficient to overcome energy losses within the tissues.

In well perfused tissues, the major loss arises from perfusion. To overcome this, SAR values in the region of $200-300\ \mathrm{W\ kg^{-1}}$ are required in the target volumes. For local hyperthermia, generators should be capable of delivering $200-500\ \mathrm{W}$, depending upon frequency of operation. For regional hyperthermia, the power requirements are typically $1500-2000\ \mathrm{W}$.

4. It is important that the impedance of the applicator, when positioned on the patient, is matched to the coaxial line and generator.

1.4 Applicators for Local Hyperthermia

1.4.1 Electric (Capacitive) Applicators

A common method of employing radiofrequency fields in local hyperthermia is to place a small electrode above the region to be treated whilst a second electrode is placed some distance from the first with the tumour between. Often, the return electrode is of larger area and is placed beneath the supine patient. The dimensions of the small electrode will depend upon the lateral dimensions of the tumour. Typically, the electrode (including its bolus) should extend beyond the tumour by $3-5$ cm in all directions. The frequencies usually used for this technique are 13.56 and 27.12 MHz although some machines designed primarily for regional hyperthermia and operating at 8 MHz can be used with a small electrode for local treatments. Thus wavelengths are much greater than

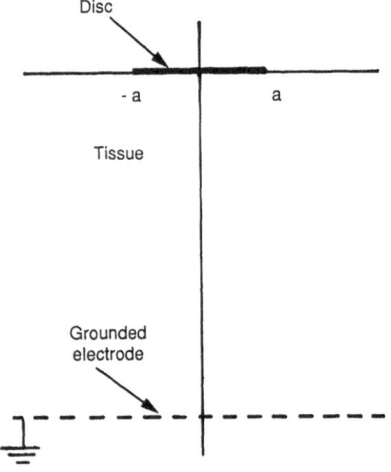

Fig. 1.12. Disc electrode

the dimensions of both applicators and treated regions and the distribution of fields and currents may be found from solving the Laplace equation.

The important characteristics of capacitive applicators are highlighted in the simple analytical model of a circular disc, radius a, in contact with a uniform medium considered by Wiley and Webster (1982) in connection with an electrosurgial technique. A grounded return electrode was assumed to be at infinity (Fig. 1.12) and the potential Φ is relative to this. The potential Φ satisfies Laplace's equation

$$\nabla^2 \Phi = 0 \qquad (1.73)$$

and the boundary conditions which must be satisfied are

$$\Phi = \Phi_0 \quad \text{at } z = 0 \quad \text{for} \quad |r| \leq a \qquad (1.74)$$

$$\frac{\partial \Phi}{\partial z} = 0 \quad \text{at } z = 0 \quad \text{for} \quad |r| > a \qquad (1.75)$$

$$\Phi \to 0 \quad \text{as} \quad |r| \to \infty \quad \text{and} \quad z \to \infty \qquad (1.76)$$

Wiley and Webster showed that

$$\Phi(r, z) = 2\Phi_0$$
$$\times \sin^{-1} \left[\frac{2a}{[(r-a)^2] + z^2)^{0.5} + [(r+a)^2 + z^2]^{0.5}} \right]$$
$$\text{for} \quad z = 0 \qquad (1.77)$$

and

$$\Phi(r, 0) = 2\Phi_0 \sin^{-1} \left[\frac{r}{a} \right] \quad \text{for} \quad z = 0 \quad \text{and} \quad |r| \geq a \qquad (1.78)$$

The electric field $\mathbf{E}(r, z)$ is found from

$$\mathbf{E}(r, z) = \nabla \Phi \qquad (1.79)$$

Figure 1.13 is a plot of $\mathbf{E}(r, z)$ and shows the large fringing fields at the edge of the disc. The absorbed power density $(\alpha |\mathbf{E}|^2)$ in the vicinity of the edges of the disc is very high but decreases rapidly with increasing distance from the disc. In practice superficial tissues must be protected from the intense fields present at distances $0 < z < 0.3\,a$ and this is achieved through use of a bolus. The need for a bolus has been indicated by practical experience. Brezovich et al. (1981) reported that flat metallic electrodes frequently produce burns in skin and subcutaneous fat, especially along the edges of the electrodes. Beyond this

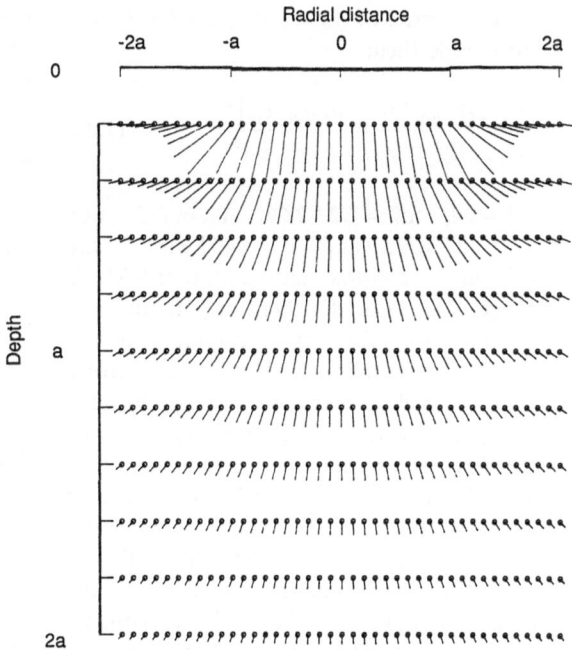

Fig. 1.13. Electric field vector beneath disc electrode

region the absorbed power density decreases with increasing z in this uniform medium with a penetration depth approximately equal to a, the radius of the disc. The assumption in this simple analysis that the second electrode is at infinity implies that in practice the separation between electrodes would be several disc diameters, limiting its application to small discs.

The second feature of Fig. 1.13 is that the electric field is predominantly perpendicular to the plane of the disc for $|r/a| < 0.7$. This is important when we consider the boundary conditions which must be satisfied in inhomogeneous media such as we might have with a capacitive applicator and bolus applied to skin and superficial fat layers overlying muscle tissue. For example, for the electric field component perpendicular to the boundary between fat and muscle the electric displacement $D(= \varepsilon E)$ is continuous. Thus:

$$\varepsilon_f E_{perp}^f = \varepsilon_m E_{perp}^m \qquad (1.80)$$

where E_{perp} is the perpendicular component of the electric field at the fat – muscle interface in the medium indicated by the superscripts. The complex permittivities ε_f and ε_m are:

$$\varepsilon_f = \varepsilon_0 \left(\varepsilon_f' - j \frac{\sigma_f}{\omega \varepsilon_0} \right) \quad \text{and}$$

$$\varepsilon_m = \varepsilon_0 \left(\varepsilon_m' - j \frac{\sigma_m}{\omega \varepsilon_0} \right) \qquad (1.81)$$

If we consider the ratio of the resulting SAR values in the fat and muscle then:

$$\frac{SAR_f}{SAR_m} = \frac{0.5\,\sigma_f E_{perp}^{f\,2}/\varrho_f}{0.5\,\sigma_m E_{perp}^{m\,2}/\varrho_m} = \frac{\varepsilon_f''\varrho_m\,|\varepsilon_m|^2}{\varepsilon_m''\varrho_f\,|\varepsilon_f|^2} \qquad (1.82)$$

Taking $\varepsilon_f' = 20$, $\varepsilon_f'' = 7.2-28.6$, $\varepsilon_m = 113$ and $\varepsilon_m'' = 399$ at 27 MHz and $\varrho_f = 940\ \mathrm{kg\,m^{-3}}$ and $\varrho_m = 1070\ \mathrm{kg\,m^{-3}}$ it follows that the ratio of SARs is in the range $7.8-11.5$. Furthermore, taking the specific heats of fat and muscle to be $3470\ \mathrm{J\,kg^{-1}\,°C^{-1}}$ and $2260\ \mathrm{J\,kg^{-1}\,°C^{-1}}$, respectively, then the rate at which temperature increases in fat is approximately $12-18$ times that in muscle. This ignores the effects of thermal conduction and blood flow, which will enhance this ratio. This problem of excessive heating of superficial fat layers is the major drawback to the technique. Although skin cooling by means of the bolus can reduce the problem, this is only useful when the thickness of the subcutaneous fat layer is not too great. The topic of skin cooling of uniformly perfused tissue has been discussed by Roemer (1988). He considered a uniformly perfused unheated tissue subject to a skin temperature T_s relative to the arterial blood temperature. The one-dimensional bioheat transfer equation for this case is:

$$\frac{d^2T}{dx^2} = \frac{w\,c_b}{k}\,T \qquad (1.83)$$

where w, c_b and k are the perfusion ($\mathrm{kg\,m^{-3}\,s^{-1}}$), specific heat of blood and thermal conductivity of the tissue. Equation (1.83) has the solution

$$T = T_s \exp\left[\frac{-x}{d_{cool}}\right] \qquad (1.84)$$

where d_{cool}, the penetration depth for the skin temperature, is

$$d_{cool} = \left[\frac{k}{w\,c_b}\right]^{0.5} \qquad (1.85)$$

The sensitivity of tissue temperature at depth x to skin temperature is

$$\frac{dT}{dx} = \exp\left[\frac{-x}{d_{cool}}\right] \qquad (1.86)$$

The result is independent of adding a heating modality. Applying this to fat (with parameters $w \approx 0.6\ \mathrm{kg\,m^{-3}\,s^{-1}}$, $k = 0.21\ \mathrm{W\,m^{-1}\,°C^{-1}}$ $c_b = 4000\ \mathrm{J\,kg^{-1}\,°C^{-1}}$), we find that $d_{cool} \approx 9$ mm. Thus cooling the skin 20 °C below arterial temperature will reduce the temperature 2 cm into a fat layer by 2 °C. Indeed, clinical experience with capacitive applicators has shown that skin cooling is ineffective in patients with fat layers thicker than $1.5-2$ cm.

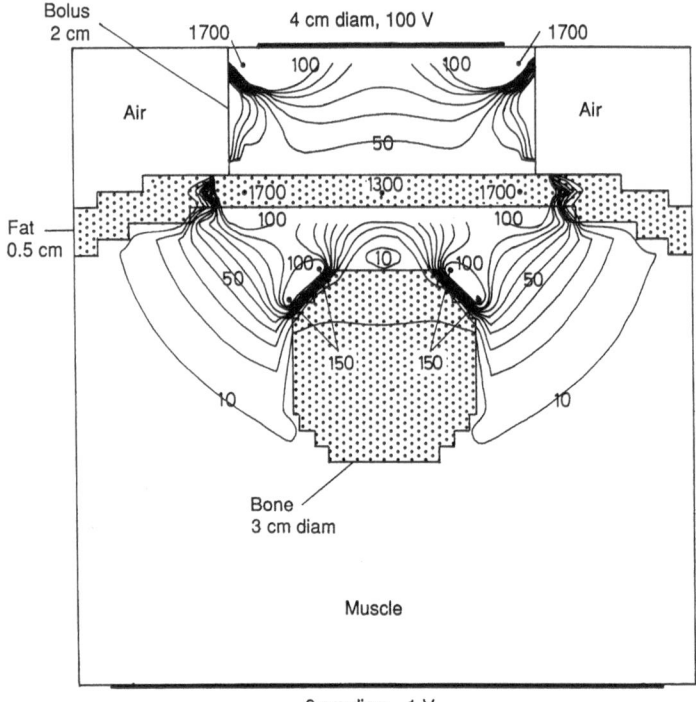

Fig. 1.14. Absorbed power density in an inhomogeneous model due to disc electrodes (Hand and Hind 1986)

Use of a bolus has other advantages. The highly unpredictable performance of electrodes placed directly on the skin has been described by Overmyer et al. (1979). In these cases non-uniform contact or perspiration may lead to regions of high current density and consequently to high temperatures. The impedance of the applicator when placed on the patient should be relatively insensitive to patient movement, obviating the need for frequent adjustment of the impedance matching network and so simplifying procedure. Bini et al. (1985) showed that the capacitive reactance of the applicator is insensitive to the thickness of the bolus provided the thickness is sufficient to contain the region of high electric field around the electrode. For applicators with electrode and bolus diameters of approximately 6 and 10 cm, respectively, the resistance presented by the tissue is typically a few tens of ohms.

The result of a two-dimensional calculation of absorbed power density within a fat-muscle-bone model is shown in Fig. 1.14. The 4 cm diameter electrode is spaced from the fat layer by a bolus and the return electrode is spaced 10 cm from the small one. The absorbed power density is shown normalised to the maximum in muscle at the fat−muscle interface. Very high absorbed power density is seen in the fat layer and the presence of the low conductivity region simulating bone disturbs the current flow, producing hot spots near to the bone and a region of reduced absorbed power density above the bone. Other examples of SAR in inhomogeneous media have been discussed (Doss 1982; Armitage et al. 1983).

A practical form of a capacitive applicator has been described by Brezovich et al. (1981). This consists of a cylindrical plastic housing with a 100 μm thick latex rubber membrane across its open end. Within the housing a brass plate acts as the electrode (Fig. 1.15).

The interior of the applicator is filled with 0.44% saline solution which has a resistivity close to that of muscle at a frequency of about 10 MHz. The solution is circulated through the applicator at a regulated temperature to give some control over skin temperature. Good electrical and thermal contact with the patient's skin is achieved through the use of conductive gel.

1.4.2 Magnetic (Inductive) Applicators

1.4.2.1 Coil Applicators

Induction coils have been used for many years in short-wave diathermy at frequencies in the industrial, scientific and medical (ISM) bands centred at 13.56 and 27.12 MHz. If the frequency is significantly lower than this, the current through the coil required to produce sufficiently high SAR in the tissues becomes impractically large.

A simple analysis of the flat "pancake coil" (Fig. 1.16) modelled as a series of concentric closed loops

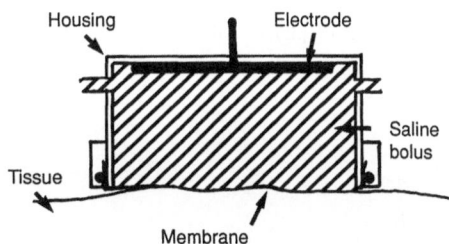

Fig. 1.15. Typical capacitive applicator for superficial hyperthermia

Fig. 1.16. Three-turn planar ("pancake") coil modelled as three concentric current loops

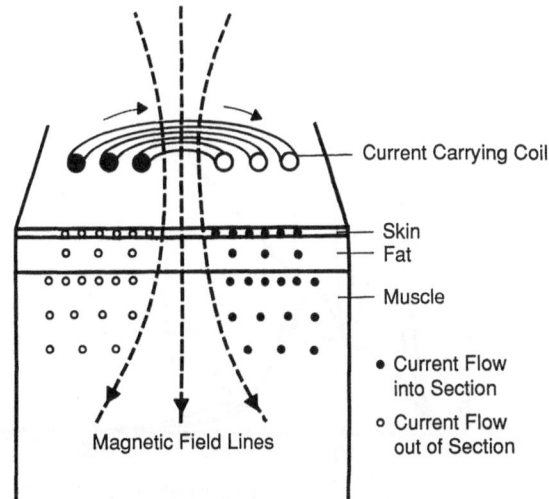

was described by Guy et al. (1974). There are two contributions to the power deposition in tissues as can be seen from the expression determining the electric field:

$$\mathbf{E} = -\nabla \Phi - \frac{\partial \mathbf{A}}{\partial t} \qquad (1.87)$$

where $\mathbf{A}$ is the magnetic vector potential. The contribution involving the gradient of the potential Φ is important close to the coil windings. Each turn of the coil is taken to be at a single potential. Since the relative permittivity of the tissues is significantly different from that of air, the potential distribution is approximated by considering the actual turns together with an image of them about the skin–air interface

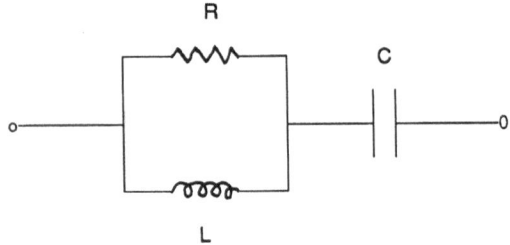

Fig. 1.17. Equivalent circuit of planar coil

rather than the series of images about each tissue interface required for a rigorous solution. The component of the electric field normal to the tissues, E_z, is found by evaluating $d\Phi/dz$. The second term is associated with the induced electric field arising from the time-varying magnetic field around the coil. The magnetically induced electric field $E_\phi = -j\omega A_\phi$. A_ϕ may be expressed in terms of complete elliptical integrals of the first and second kinds. When the coil is suitably spaced from the tissue, the dominant component is the rotational E_ϕ.

The pancake coil is an example of an inductive applicator which can be considered in terms of a magnetic dipole moment orientated perpendicularly to the skin surface (Morita and Bach Andersen 1982). Where E_ϕ is dominant, the induced currents in the tissue form closed loops (eddy currents) and from symmetry the SAR on the central axis of the coil is zero. This distribution of SAR is a drawback with this class of applicator. Since E_ϕ is parallel to the skin and fat-muscle interface, the boundary conditions require that the field be continuous across the interface, i.e.

$$E_\phi^f = E_\phi^m \qquad (1.88)$$

and so, taking $\sigma_f = 0.011 - 0.043 \, \text{S m}^{-1}$ and $\sigma_m = 0.6 \, \text{S m}^{-1}$, the ratio of SAR values across the fat-muscle interface is

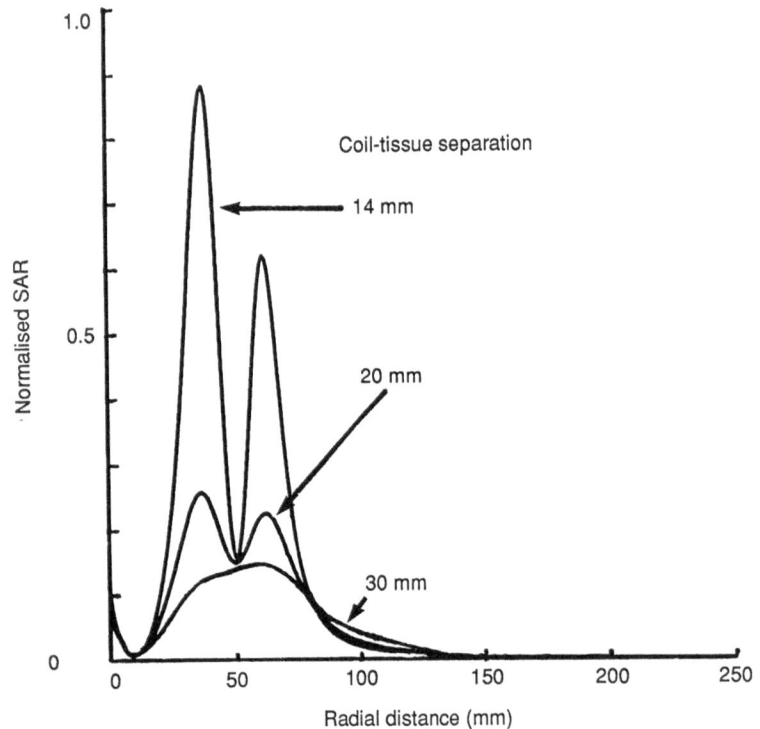

Fig. 1.18. Normalised SAR at surface of layered tissue due to three-turn planar coil as function of coil-tissue separation (Hand and Hind 1986)

$$\frac{SAR_f}{SAR_m} = \frac{\sigma_f \varrho_m}{\sigma_m \varrho_f} \approx 0.02 - 0.08 \tag{1.89}$$

in marked contrast with the case of the electric applicator.

The impedance of the equivalent circuit shown in Fig. 1.17 is:

$$Z_e = \frac{R}{\omega^2 C^2 R^2 + (1-\omega^2 LC)^2}$$
$$+ j \frac{\omega L(1-\omega^2 LC) - CR^2}{\omega^2 C^2 R^2 + (1-\omega^2 LC)^2} \tag{1.90}$$

or $\quad Z_e = R_e + j\omega L_e \tag{1.91}$

where R_e and L_e are the effective resistance and inductance at frequency $f = \omega/2\pi$. Except when $\omega \approx (LC)^{-0.5}$, we can take L_e as:

$$L_e \approx \frac{L}{(1-\omega^2 LC)} \tag{1.92}$$

When $\omega^2 LC > 1$, $L_e < 0$ and the applicator will show characteristics of a capacitive device. To avoid this the coil should be restricted to a small number of adequately spaced turns and used at frequencies well below its resonant frequency.

Figure 1.18 shows the effect of varying the spacing between tissues and a three-turn coil (with radii of 4, 5 and 6 cm) driven at 27 MHz on the absorbed power density at the surface of a 1 cm thick layer of fat overlying muscle tissue. When the separation between coil and tissues is less than about 3 cm, the E_z component due to the potential gradients around the turns dominates, leading to high absorbed power density in the fat. For larger separations, the E_ϕ component dominates and the absorbed power density in the muscle shows an effective penetration depth of 3–4 cm, the larger value being associated with larger coils. The effects of coil size, fat thickness and skin cooling on temperature distributions have been predicted by Hand et al. (1982). Further discussion of coil applicators is to be found in Lehmann et al. (1968, 1969), Lerch and Kohn (1983) and Kantor and Moon (1983).

1.4.2.2 Distributed Current Applicators

Another class of magnetic applicators are those in which the loop of current-carrying conductors is in a plane perpendicular to the skin surface, i.e. with a conductor parallel to the skin and the conductor carrying the return current behind the primary conductor. Such applicators may be considered in terms of a magnetic dipole moment parallel to the body surface (Morita and Bach Andersen 1982) and examples include the distributed current device described by Bach Andersen et al. (1984), the twin dipole applicator described by Franconi et al. (1986) and the distributed current device described by Johnson (1986).

The applicator described by Bach Andersen et al. (1984) is based on a transmission line formed by a current-carrying strip approximately 15 mm wide placed about 15 mm from a reflector. The line is tuned for resonance by an external capacitor at one end and short-circuited at the other end (Fig. 1.19). Particular dimensions for the applicator may be accommodated by selecting the length of the line, L (6–10 cm), and placing several lines in parallel to give a width in the range 4–9 cm. The operating frequency is around 150 MHz. There is a component of electric field perpendicular to the skin associated with the transmission line, so use of a bolus is advisable. The device is narrow band.

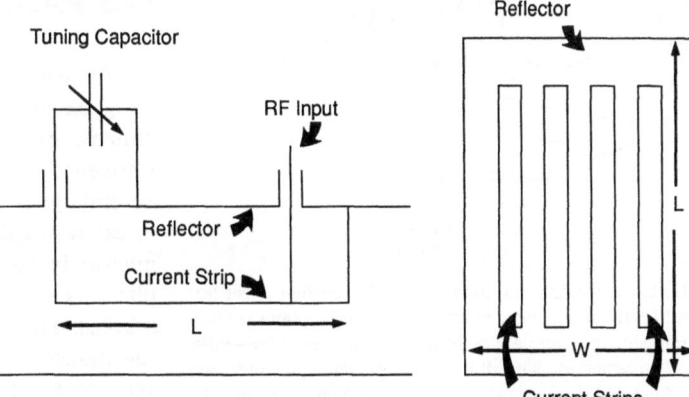

Fig. 1.19. Current strip applicator (after Bach Andersen et al. 1984)

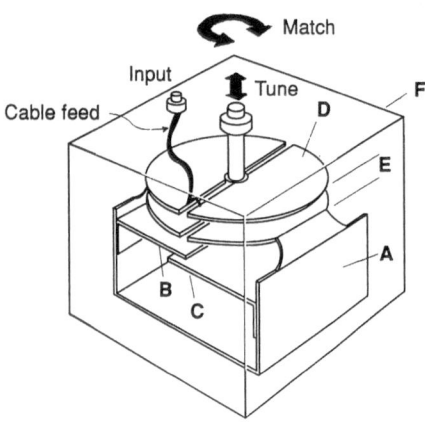

Fig. 1.20. Current sheet applicator (after Johnson 1986). The "U"-shaped high conductivity plate *A* has capacitor plates *B* and *C* added. Power is coupled through the split disc *D*. The resonant frequency of the applicator is adjusted by changing the distance *E* between disc and resonator. The complete structure is housed in a screening box *F*

The applicator described by Johnson (1986) consists of a flat high conductivity plate folded to form a "U" with capacitor plates between the arms of the "U". This resonant circuit is placed in a screening box which has an aperture covered by low loss dielectric material (Fig. 1.20). The resonant frequency is determined by the capacitance between the arms of the "U" and by the area of the loop formed by the "U" and the capacitor plates. It is selected by altering either the distance between the plates and the lower surface of the "U" (the "radiating" surface) or the

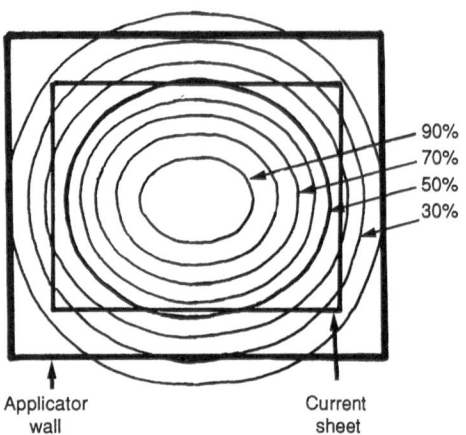

Fig. 1.21. SAR distribution due to 434-MHz current sheet applicator. Measurements were made 1 cm from the plane surface of a homogeneous muscle phantom. Contours are 20% – 90% (in 10% increments). The dimensions of the applicator are 7.2 cm × 6 cm and those of the current sheet 5.5 cm × 4.9 cm. The direction of current flow is parallel to the longer dimensions

number of plates and spacing between them. By maintaining the widths of the radiating surface and capacitor plates at a constant ratio, the width of the former can be chosen independent of resonant frequency. Applicators using this concept have been built to operate at frequencies from 20 MHz to 915 MHz. At frequencies greater than about 100 MHz, RF power is coupled to the resonant structure by means of a split disc. A good impedance match to a 50-ohm coaxial line is achieved by rotating the disc whilst small changes to the resonant frequency may be made by adjusting the separation of disc and resonant circuit. At lower frequencies, coupling is by means of a loop antenna. This type of applicator is also narrow band.

The SAR distribution for a current sheet applicator measured in a plane 1 cm from the surface of a homogeneous muscle phantom (Chou et al. 1984) is shown in Fig. 1.21. The peak SAR measured occurred close to the centre of the radiating surface and the 50% SAR contour extends almost to the edge of this surface. The 25% SAR contour extends approximately to the aperture of the screening box. This measured distribution is in close agreement with predicted values (Johnson et al. 1987).

A distributed current on a conducting sheet is the basis of an applicator described by Kato and Ishida (1983). This is a one-turn, square column-like coil driven at 6 MHz with the plane of the current-carrying loop perpendicular to the tissue surface. The lateral dimensions of the applicator described were 20 cm × 20 cm and its depth was 60 cm but since the resonant frequency of the device is determined by its inductance and capacitance, its size can be chosen to meet specific requirements. Furukawa et al. (1988) describe its use in treating patients with tumours at depths of up to 5 cm.

1.4.3 Radiating Applicators

At frequencies above about 200 MHz applicators designed for local hyperthermal treatments have dimensions which are comparable with the wavelength and so radiation efficiency is increased. Several types of applicator based on hollow cylindrical waveguides with various cross-sections or on microstrip have been developed. Most radiative applicators are designed to be used in direct contact with a bolus. This improves the coupling of the E-field to the tissues, avoids exposing the tissues to the fields very close to the aperture of the applicator and maintains leakage fields at an acceptably low level.

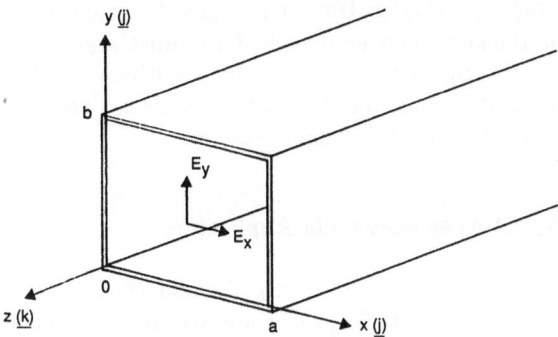

Fig. 1.22. Coordinate system for waveguide applicator with rectangular cross-section (Hand and Hind 1986)

1.4.3.1 Rectangular Waveguide Applicators

Many radiative applicators are based on waveguides with rectangular cross-section (Fig. 1.22) operated in a TE mode. The electric field in this structure is found by solving Maxwell's equations subject to the boundary conditions which must be satisfied at the waveguide walls, which are taken to be perfect conductors. The electric field components are (Johnk 1975):

$$E_x = \frac{j\omega\mu}{h^2}\frac{n\pi}{b} A \cos\left(\frac{m\pi x}{a}\right)\sin\left(\frac{n\pi y}{b}\right) \qquad (1.93)$$

$$E_y = -\frac{j\omega\mu}{h^2}\frac{m\pi}{a} A \sin\left(\frac{m\pi x}{a}\right)\cos\left(\frac{n\pi y}{b}\right) \qquad (1.94)$$

$$h^2 = \gamma_{mn}^2 + \omega^2\mu\varepsilon = \left(\frac{m\pi}{a}\right)^2 + \left(\frac{n\pi}{b}\right)^2 \quad \text{and}$$

$$A = \text{constant.}$$

m and n are integers (but not both equal to zero) and define the number of half sine variations in the x- and y-directions, respectively, exhibited by the field. The medium within the applicator is assumed to have permittivity ε and permeability μ. The propagation constant γ_{mn} is:

$$\gamma_{mn} = \left[\left(\frac{m\pi}{a}\right)^2 + \left(\frac{n\pi}{b}\right)^2 - \omega^2\mu\varepsilon\right]^{0.5} \qquad (1.95)$$

From Eq. (1.95), $\gamma_{mn} = j\beta_{mn}$ (i.e. a phase constant) if $\omega > \omega_{cmn}$ where ω_{cmn} is:

$$\omega_{cmn} = \left\{\frac{1}{\mu\varepsilon}\left[\left(\frac{m\pi}{a}\right)^2 + \left(\frac{n\pi}{b}\right)^2\right]\right\}^{0.5} \qquad (1.96)$$

If $\omega \leq \omega_{cmn}$ then $\gamma_{mn} = \alpha_{mn}$, i.e. a real attenuation constant. The frequency $f_c = \omega_{cmn}/(2\pi)$ is the cut-off frequency for the TE_{mn} mode.

For cross-sections with a > b the lowest cut-off frequency, which occurs for the TE_{10} mode, is:

$$f_{cTE_{10}} = \frac{1}{2a[\mu\varepsilon]^{0.5}} \qquad (1.97)$$

Operation in the TE_{10} mode can be ensured by choosing a loading material with suitable ε (assuming

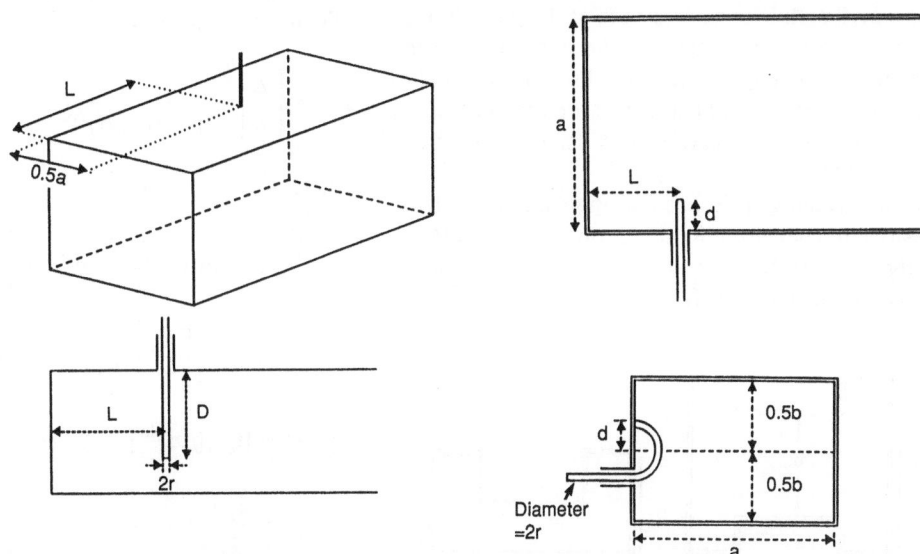

Fig. 1.23. Probe and loop antennas for coupling power from the feed line to the applicator. Details of determining the critical dimensions a, r, D, L, d and b are given in Collin (1960) and Johnson (1965)

$\mu \approx \mu_0$) and dimensions for the applicator such that the operating frequency f is approximately $10\% - 30\%$ higher than F_{cTE10}. Low loss (tan $\delta \approx 10^{-3} - 10^{-4}$) dielectric materials ($\varepsilon'$ in the range 2-140) which are suitable for this purpose may be obtained commercially in several forms, e.g. solid, powder, foam sheet or castable materials. Details of some of these have been listed by Hand and Hind (1986).

The techniques of coupling power from the feed line to the applicator by means of a probe or loop antenna (Fig. 1.23) are discussed in detail by Collin (1960) and Johnson (1965). By choosing the dimensions of the coupling antenna and its position with respect to the closed end of the applicator, the radiation resistance of the probe or loop can be made equal to the characteristic impedance of the coaxial line and the sum of the input reactances of the TE_{10} mode and all higher modes can be made zero. It is also important that the applicator is sufficiently long that the aperture is about a quarter wavelength in the applicator or more from the coupling antenna.

The characteristic impedance Z_{TE10} of the applicator is:

$$Z_{TE10} = \left\{ \frac{\mu}{\varepsilon} \left[1 - \left(\frac{f_{cTE10}}{f} \right)^2 \right]^{-1} \right\}^{0.5} \qquad (1.98)$$

For $f/f_{cTE10} \approx 1.3$, this is approximately 240 ohms for an air-filled applicator and is higher than the characteristic impedance of tissues.

An improved match should be achieved with dielectrically loaded applicators. If a section which provides an impedance equal to the geometric mean of the impedances of the applicator and load and is a quarter of a wavelength long as measured at the phase velocity in this region is placed at the aperture, the impedance match may by improved. Final adjustment of the impedance of applicator plus load to that of the coaxial line may be carried out using a stub tuner (Sect. 1.3.5).

It is an advantage if impedance matching is fairly insensitive to changes in load. Antolini et al. (1984) investigated the effect of water bolus thicknesses on the impedance match to different tissue loads and showed

that the reflection coefficient Γ is less than 0.05 for bolus thickness greater than half the wavelength in water. A compromise is usually required because of the decreased coupling of power to the tissues and a bolus $2 - 3$ cm thick is usually chosen.

1.4.3.2 Ridged Waveguide Applicators

Waveguides with a ridged cross-section (Fig. 1.24) have a lower cut-off frequency, a lower characteristic impedance and a wider frequency band free from higher order mode interference compared with regular rectangular waveguides of the same dimensions. The cut-off frequency for the dominant TE_{10} mode is (Chen 1957):

$$f_{c10} = \frac{1}{\pi (\mu \varepsilon)^{0.5} \left[((a_2/b_2) + Q)(a_1 - a_2)b_1 \right]^{0.5}} \qquad (1.99)$$

where

$$Q = \frac{2}{\pi} \left\{ \frac{x^2 + 1}{x} \cosh^{-1} \left[\frac{1 + x^2}{1 - x^2} \right] - 2 \ln \left[\frac{4x}{1 - x^2} \right] \right\}$$

and $x = \dfrac{b_2}{b_1}$

The ratio a_2/a_1 should be in the range $0.4 - 0.5$ for the greatest reduction in f_{c10} and in the range $0.3 - 0.5$ for maximum separation from higher order modes (Hopfer 1955). The aspect ratio b_1/a_1 should be approximately 0.4.

The characteristic impedance of a single-ridged applicator at frequency f is:

$$Z_{10} = \left[\frac{\mu}{\varepsilon} \right]^{0.5} \frac{S}{[1 - (f_{c10}/f)^2]^{0.5}} \qquad (1.100)$$

where

$$S = \left\{ Q \cos \psi_2 + \frac{\lambda_c'}{b_2 \pi} \left(\sin \psi_2 + \left[\frac{b_2}{b_1} \right] \right. \right.$$
$$\left. \left. \cos \psi_2 \tan \left[\frac{\psi_1}{2} \right] \right) \right\}^{-1} \quad \text{with}$$

$$\lambda_c' = [f_{c10} [\mu \varepsilon]^{0.5}]^{-1} \ , \quad \psi_1 = \left[1 - \frac{a_2}{a_1} \right] \frac{a_1 \pi}{\lambda_c'} \ ,$$

$$\psi_2 = \frac{a_2 \pi}{\lambda_c'}$$

and f_{c10} and Q are the same as in Eq. (1.99).

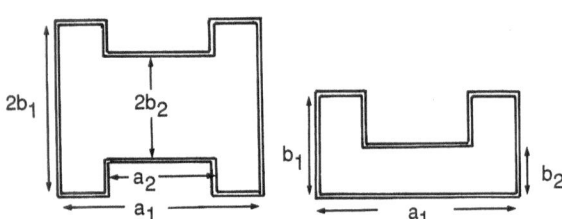

Fig. 1.24. Applicators with ridged cross-sections

The characteristic impedance of a dual-ridged waveguide is $2Z_{10}$. Hand and Hind (1986) list characteristic impedances for ridged applicators of typical dimensions and loaded with a range of dielectric materials. Although they achieve a better impedance match to tissues than regular rectangular applicators, ridged applicators do have a drawback in that the effective aperture is approximately that region beneath the ridges. This can be appreciably smaller than that of a regular rectangular applicator of similar overall dimensions, especially if low values of f_{c10} and Z_{10} are required. Details of a dielectrically loaded dual-ridged applicator for local treatments are given by Turner (1983).

1.4.3.3 Dielectric Slab-Loaded Rectangular Applicator

A method of improving the uniformity of the E-field across the aperture of a rectangular waveguide applicator is to partially load it with dielectric slabs (Fig. 1.25). Hudson (1957) showed that this structure can support a TEM wave in the central region at a frequency f_{TEM} given by:

$$f_{TEM} = \frac{c}{4s\left[(\varepsilon_1' - \varepsilon_2')/\varepsilon_2'\right]^{0.5}} \qquad (1.101)$$

At this frequency the electric field is in the y-direction and is uniform across the central region. To avoid higher order modes the conditions

$$b < \frac{\lambda}{2} \quad \text{and}$$

$$d < \frac{\lambda}{\pi} \tan^{-1}\left\{ -\left[\frac{\varepsilon_1'}{\varepsilon_2'}\right]^{0.5} \tan\left(\frac{\pi}{2}\left[\frac{\varepsilon_1'}{\varepsilon_1' - \varepsilon_2'}\right]^{0.5}\right)\right\}$$

where λ is the wavelength of a plane wave in an unbounded region with relative permittivity ε_2' (van Koughnett and Wyslouzil 1972) must be satisfied.

This type of applicator is usually excited by a probe antenna parallel to the walls of the slabs and positioned symmetrically in the central region. The TEM wave is simulated in the central region only at f_{TEM}, and in practice operating at a driving frequency which deviates from the TEM condition results in an electric field distribution with a maximum or local minimum in the central region according to whether $f < f_{TEM}$ or $f > f_{TEM}$, respectively. In practice, this type of applicator can be expected to provide a variation in E_y^2 across the central region of about 20% due to tolerances in construction and materials (Hand and Hind 1986). Descriptions of slab-loaded applicators appear in Cheung et al. (1977) and Kantor and Witters (1980).

1.4.3.4 Microstrip and Other Compact Applicators

Although waveguide applicators are generally very useful in treating superficial tumours, there are cases in which their bulk, even when dielectrically loaded and/or ridged, makes access to the tumour difficult. To improve these situations, applicators based on microstrip or similar structures have been developed. A microstrip transmission line consists of a thin substrate of dielectric material of relative permittivity ε' (usually $\varepsilon' < 10$) with a metallic ground plane on one side and a metallic line on the other side (Fig. 1.26). The lowest order mode which can be propagated along the line closely resembles a TEM mode and has no cut-off frequency. The wavelength λ_m is:

$$\lambda_m = \frac{\lambda_0}{\varepsilon_e'^{0.5}} \qquad (1.102)$$

where λ_0 is the free-space wavelength and ε_e' is an effective relative permittivity which takes the fringing

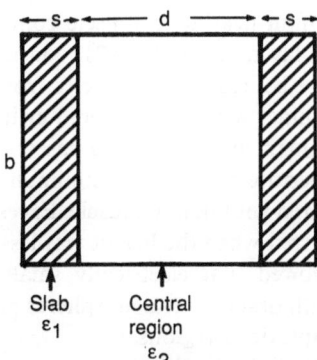

Fig. 1.25. Slab-loaded applicator with rectangular cross-section

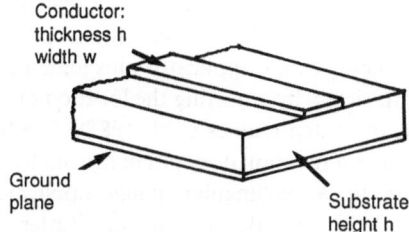

Fig. 1.26. Microstrip transmission line

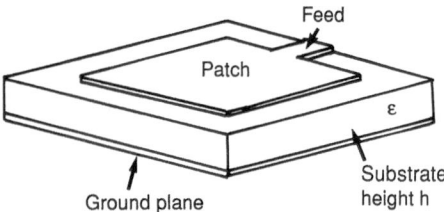

Fig. 1.27. Microstrip rectangular patch applicator

fields around the line into account. This effective relative permittivity is less than the relative permittivity of the substrate and is a function of ε', the thickness of the substrate, h, and the width of the line, w. Methods for determining ε_e' are described by James et al. (1981).

A simple microstrip antenna consists of a rectangular metallic patch approximately $0.5\,\lambda_m$ long on the substrate (Fig. 1.27). However, the fact that applicators for hyperthermia must operate in close proximity to the tissues presents a problem since the characteristics of microstrip antennas are sensitive to the dielectric properties of the media close to them. Estimates of the changes for rectangular patch and ring antennas have been discussed by Bahl and Stuchly (1980), Bahl et al. (1980, 1982) and Sandhu and Kolozsvary (1984). These authors used a variational method similar to that developed by Yamashita and Mittra (1968) to calculate line capacitance and to find the effective dielectric constant and conductivity of the microstrip antenna plus its layered dielectric load in order to determine the resonant frequency for various loads.

For a sufficiently thick fat layer (greater than about 2 cm) the effective permittivity of the applicator plus load and the resonant frequency of the applicator become insensitive to changes in fat thickness. In practice the provision of a water bolus 2 cm or more thick should result in a resonant frequency which does not vary significantly with load. The bolus also prevents exposing the tissues to the relatively high fields which occur at the edge of the patch.

Microstrip patch applicators may be excited by either a microstrip line or a coaxial line with its central conductor attached to the metallic patch and its outer conductor attached to the ground plane. Impedance matching of the applicator plus load may be achieved empirically by adjusting the location of the feed point on the patch (James et al. 1981). The SAR distribution as a function of depth in layered biological media due to a rectangular patch applicator has been modelled by Beyne and de Zutter (1988). This rigorous model took account of the surface current

distribution near the edges of the patch and excitation of the patch by a coaxial feed.

Tanabe et al. (1983) described a microstrip spiral applicator for use at 915 MHz. The 4-cm Archimedean spiral on a substrate with $\varepsilon' = 2.3$ gives the applicator a broader bandwidth than a rectangular or circular patch and a circularly polarised electric field. Other spiral applicators, including an 8 cm diameter device for use at 434 MHz, are described by Kapp et al. (1988). The heating patterns of these applicators are well approximated by Gaussian functions with 50% widths of 2.8 cm and 7.0 cm, respectively. Recently Ryan et al. (1988) have investigated the use of three types of spiral and applicators with a bolus of fixed thickness for use at 433 MHz.

A range of applicators related to microstrip patch antennas has been developed by Johnson et al. (1984). These applicators differ from other patch-based devices in that they have a substrate of high relative permittivity ($\varepsilon' \approx 30$) and a cover of the same material over the patch. This structure improves power transfer to the tissues and reduces the sensitivity of the resonant frequency and impedance to variations in loading conditions. James et al. (1986) describe techniques for designing low frequency applicators for local treatments, e.g. by using a $\lambda_m/4$ patch shorted to the ground plane at one edge and/or magnetic material with an effective relative permeability $\mu_r > 1$.

1.4.4 Applicator Size and Penetration Depth

Guy (1971) evaluated the electromagnetic fields and associated SAR due to a rectangular aperture source at 915 MHz in direct contact with plane layers of fat and muscle using a Fourier transform technique (Harrington 1961). Hot spots near the edge of the applicator due to the rapidly diverging fields at these discontinuities were clearly identified. Guy concluded that a TE_{10} mode distribution on an aperture λ_f high and between λ_f and $2\lambda_f$ wide should produce optimum heating in muscle compared with that in the fat layer. Ho et al. (1971) considered TE_{10} mode aperture sources at frequencies from 433 to 2450 MHz in contact with triple-layered cylinders which simulated human limbs. In these cases also, the height of the aperture was λ_f for minimum heating in fat compared with that in muscle. Excessive heating of fat occurred when the height was less than $\lambda_f/2$. Ho (1979) showed that electrically small apertures in contact with tissue equivalent spheres produced a high SAR in superficial regions.

The field distributions on apertures of hyperthermia applicators are often approximated by assuming that

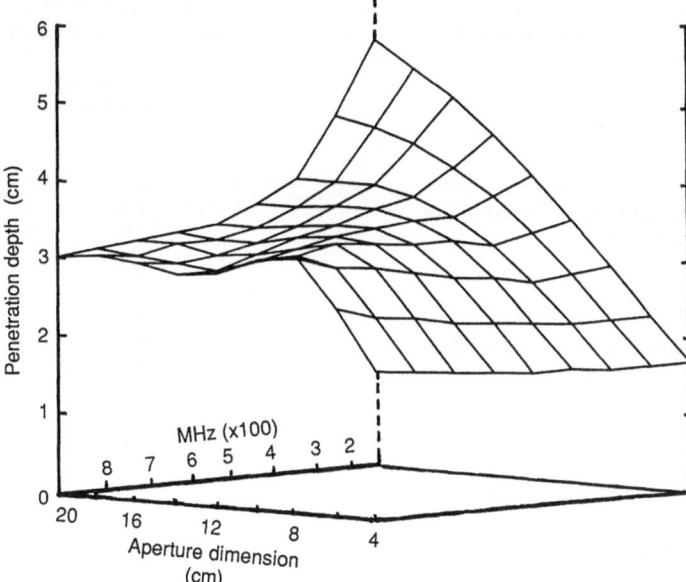

Fig. 1.28. Penetration depth (e^{-2}) in homogeneous muscle phantom due to square aperture sources as functions of aperture dimension and frequency

the aperture is on a perfectly conducting plane in contact with the tissues. Turner and Kumar (1982) used the field equivalence theorem (Jordan and Balmain 1968) to model aperture sources and assumed that the field distribution across the aperture was the ideal TE_{10} one. Better approximations to the fields in layered media in which coupling between the applicator and tissue is taken into account by considering both incident and scattered fields have been reported by Bozzetti et al. (1983) and Edenhofer (1983). Turner and Kumar (1982) showed that effective penetration depths associated with apertures typical of hyperthermia applicators were less than those associated with plane waves. Using a similar method, Hand (1987) calculated penetration depths (e^{-2}) for square apertures (4×4 cm to 20×20 cm) driven at frequencies from $100-900$ MHz assuming the aperture was in direct contact with the plane surface of an muscle-like medium (Fig. 1.28). Effective penetration depths for rectangular TE_{10} mode applicators have also been considered by Audet et al. (1980), Robillard et al. (1980), Plancot (1983), Nilsson (1984) and Cheever et al. (1987). Effective penetration depths in high water content tissue for several types of applicator covering a wide range of frequencies are listed in Table 1.12 and compared with the corresponding plane wave penetration depth.

The characteristic of an applicator referred to as the penetration depth appears frequently in the literature but its meaning requires comment. In the engineering literature penetration depth, d, is usually taken to be the distance into a phantom or tissue over which the electric field of an applicator is reduced by a factor e

[corresponding to a reduction in SAR by a factor e^2 (i.e. reduced to $\approx 13\%$)]. In the clinical literature penetration depth often refers to $d_{1/2}$, the distance over which SAR is reduced by a factor of 2, since this higher SAR level appears to be of greater relevance when using an applicator on patients. The relationship between the two parameters for plane waves is:

$$d_{1/2} = 0.346 \, d \qquad (1.103)$$

Far from an applicator the (e^{-2}) penetration depth may be estimated using

$$d = \frac{1.414c}{\omega[(\varepsilon'^2 + \varepsilon''^2)^{0.5} - \varepsilon']^{0.5}} \qquad (1.104)$$

where c = velocity of light, since d can be taken as the plane wave penetration depth. Therefore, if SAR is measurd at two depths, z_1 and z_2, beneath the centre of an aperture,

$$\frac{\text{SAR at } z_2}{\text{SAR at } z_1} = \exp\left[-2\frac{(z_2 - z_1)}{d}\right] \qquad (1.105)$$

and for a given difference between z_1 and z_2 this ratio is independent of the actual depth if z_1 and z_2 are roughly equal to or greater than the wavelength in tissue. In this case the determination of d is clearly defined. However, most hyperthermia applicators are used in such a way that superficial tissues are located at a distance less than the wavelength from the aperture. In these cases the distance d_{eff} over which the

Table 1.12. Dependence of effective penetration depth (e^{-2}) on applicator type, dimensions and operating frequency. Plane wave penetration depths and wavelengths are included for comparison. The medium is assumed to be homogeneous muscle

Applicator	Frequency (MHz)	Penetration Depth (e^{-2}) (cm)		Wavelength (cm)
		Effective	Plane wave	
Radiating aperture 10 cm × 10 cm	915	3.0	3.1	4.5
(Hand 1987)	433	3.1	3.6	8.8
	300	3.1	3.9	11.8
	200	3.1	4.8	16.6
	100	3.2	6.7	27.0
Raditating aperture 20 cm × 20 cm	915	3.0	3.1	4.5
(Hand 1987)	433	3.3	3.6	8.8
	300	3.8	3.9	11.8
	200	4.3	4.8	16.6
	100	5.4	6.7	27.0
Capacitive electrode 5 cm diameter (Hand and Johnson 1986)	27	3.0	14.3	68.1
Inductive planar coil 3 turns, 12 cm diameter (Hand and Johnson 1986)	27	3.0	14.3	68.1
Ridged waveguide effective aperture 29 cm × 13.7 cm (van Rhoon et al. 1984)	27	5.9	14.3	68.1
Inductive distributed current device aperture 20 cm × 22 cm (Johnson et al. 1987)	27	6.7	14.3	68.1

SAR is reduced by a factor e^2 is less than d and is a function of z, i.e. the penetration depth is not uniquely defined. If the performance of applicators is to be compared or assessed, the method of determining d_{eff} must be defined. The Technical Committee of the European Society for Hyperthermic Oncology defined ($d_{1/2}$) penetration depth in terms of SAR measured in a plane homogeneous muscle phantom as the distance below 5 mm at which the SAR is 50% of the SAR at the 5 mm depth (Hand et al. 1989b). Since the presence of a bolus will increase the distance from the aperture and the position at which measurements are made, the penetration depth must be determined with the bolus present, if appropriate. In the same document, this committee defined the effective field size of an applicator as the contour of 50% peak SAR measured at a depth of 5 mm below the surface of the muscle phantom.

1.4.5 Arrays of Applicators

Some of the limitations of using a single stationary applicator of any type and the potential advantages to be gained by using arrays of applicators were outlined in Sect. 1.3.2. In this section advances in the development of such arrays will be discussed.

1.4.5.1 Coherent and Incoherent Systems

When using a system with more than one applicator a choice must be made between incoherent and coherent operation, as was mentioned in Sect. 1.3.4. To illustrate the differences which may arise between the SAR distributions produced, consider a point P in a region of tissue at which significant SAR can be produced by either of two applicators. If we assume that the electric field at P due to applicator A is $\mathbf{E}_A$ and that due to applicator B is $\mathbf{E}_B$ and, for simplicity, that $\mathbf{E}_A$ and $\mathbf{E}_B$ have the same polarisation, then the following possibilities for the SAR at P arise.

$$\text{Applicator A only: SAR} = \frac{\sigma}{2\varrho} |\mathbf{E}_A|^2 \qquad (1.106)$$

$$\text{Applicator B only: SAR} = \frac{\sigma}{2\varrho} |\mathbf{E}_B|^2 \qquad (1.107)$$

A and B driven incoherently:

$$\text{SAR} = \frac{\sigma}{2\varrho} [|\mathbf{E}_A|^2 + |\mathbf{E}_B|^2] \qquad (1.108)$$

This would also be the time-averaged SAR if the applicators were driven sequentially.

A and B driven coherently:

$$\text{SAR} = \frac{\sigma}{2\varrho}\left[|\mathbf{E_A}+\mathbf{E_B}|^2\right] \qquad (1.109)$$

Assuming that the fields are sinusoidal functions of time, we have $\mathbf{E_A}(t) = \mathbf{E_A}\exp[j(\omega t+\phi_A)]$ and $\mathbf{E_B}(t) = \mathbf{E_B}\exp[j(\omega t+\phi_B)]$ and so the resultant of the vector sum in Eq. (1.109) is dependent upon the relative phase angle $\phi = \phi_A-\phi_B$ between the fields from the two applicators. For example, if $|\mathbf{E_A}| = |\mathbf{E_B}| = E$ then the SAR at P for the coherent case ranges from 0 to $2\sigma E^2/\varrho$ depending upon the value of ϕ compared with $\sigma E^2/\varrho$ for the case of incoherent excitation. This simple argument can be extended to the case of N applicators in which case there is an N-fold increase in SAR at a point where the contribution from each applicator is equal in magnitude, has the same phase and the same polarisation. Clearly constructive interference which leads to higher SAR in some parts of the field is associated with destructive interference and lower SAR in other parts of the field.

Plancot (1983) and Nilsson et al. (1985) calculated the absorbed power distribution in a homogeneous medium due to two TE_{10} mode aperture sources as a function of their relative orientation and phase and showed that some improvement in the size and uniformity of the distribution is achieved relative to the case of incoherent excitation.

1.4.5.2 Phased Arrays

The use of phased arrays of microwave antennas in free space is well established in radar and telecommunications applications, and the possible advantages of applying this approach to microwave hyperthermia include increasing the absorbed power in target regions and modifying the absorbed power distribution during a treatment. However, evaluating the potential of this approach in hyperthermia is complicated by the lossy and inhomogeneous nature of tissues. A problem in implementing the technique is to determine the relative phases and amplitudes needed for excitation of the array. An approach to this problem is to use conjugate phase with or without amplitude inversion (Skolnik and King 1964; Bach Andersen 1985; Guo et al. 1984; Loane et al. 1986). In this method a small invasive probe antenna is placed in the target volume and used to radiate a signal which is received by each element in the array applicator. If the voltage received at the ith element (i = 1 to N) is $V_i = A_i\exp[-j\phi_i]$ then by driving the elements with

a voltage proportional to $A_i\exp[j\phi_i]$ the field produced by the array will be focussed at the position the antenna was located at in the target volume. This implies that those elements more distant from the target volume should be driven with less power to minimise power absorption in superficial regions relative to that in the target volume. Alternatively, the small invasive antenna can be used to receive signals from each of the array elements in turn, enabling the relative phase and amplitude conditions for a focus at the antenna's position to be determined. Other relative excitations may be chosen, for example one which results in each element producing a signal of equal amplitude at the focus. This gives maximum resolution (a better defined focus) but at the expense of greater absorbed power beneath distant elements. Such an excitation might be appropriate if the array is remote from the tissue surface, as in the studies of Guo et al. (1984). Gee et al. (1984) calculated the power deposition in a plane, semi-infinite muscle phantom ($\varepsilon = 47-j16.2$) due to a close packed hexagonal planar array of 19 antennas (2450 MHz), offset from the tissue by a region of deionised water approximately 7 λ thick. This array produced a focus at a depth of 2.5 cm in the phantom. The half-power width of the focus was approximately one wavelength in muscle. Loane et al. (1986) considered a seven-element planar array operating at 1500 MHz and designed to be operated in water ($\varepsilon \approx 78-j19.5$). In this report the focal plane was 6 cm (2.7 λ) from the plane of the apertures and the power received from when the seven elements were focussed was 9 dB higher than that received from a single element (which was ≈ -17 dB with respect to the power at the aperture plane). These authors also demonstrated that the array could be focussed in the presence of a dielectric scatterer.

The performance of planar 4×4 arrays of small aperture sources (434 and 915 MHz) in contact with a plane muscle phantom ($\varepsilon = 51-j25$ at 915 MHz; $\varepsilon = 53-j49$ at 434 MHz) was calculated by Hand et al. (1986). The SAR distribution peaked sharply beneath each aperture and there was a rapid decrease in peak values with increasing distance into the phantom, suggesting that in practice the array would require spacing from the tissue by a bolus several cm thick. When focussed at depths greater than about 4 cm the 434 MHz array produced a small improvement in penetration compared with a single aperture source of the same dimensions as the complete array. It was also predicted that the 915-MHz array could increase the absorbed power at 2 cm into the tissue by 1.5 dB when the plane of the array was 4 cm from the phantom surface. The half-power widths of the foci were $0.7-1.1\ \lambda$ depending upon the relative amplitudes allocated to each source. Of the two frequencies considered, 434 MHz appeared to be a better choice in that greater penetration and, due to the longer wavelength, a smoother SAR distribution could be achieved. Use of the lower frequency would also decrease reflections from fat layers (Johnson et al. 1984) as well as the sensitivity of the phase conditions for focussing to inhomogeneities and tissue movement. A related study (Hand et al. 1985) suggested that no advantage is to be gained by operating at 300 MHz. At the frequencies considered in these studies a variation in phase of $\pm \pi/5$ about the desired values produced insignificant changes in the SAR distribution in the homogeneous medium.

Convergent lens type waveguide applicators operating at 2450 MHz and 434 MHz have been described by Nikawa et al. (1985) and Takahashi et al. (1985). In this type of device, metal plates are placed in the waveguide parallel to the short walls and divide the waveguide near the aperture into four zones. The lengths of the plates, their dependence on height in the waveguide and their separation determine the focussing characteristics in both E- and H-planes. A modest increase in penetration over that for a simple waveguide applicator was reported but the peak SAR remains close to the applicator due to the high attenuation in the medium.

The studies mentioned above indicate that planar arrays can produce a local maximum in the focal plane but that the high attenuation associated with tissues of high water content results in field levels in the focal plane that are considerably lower than those in more superficial tissues. For this reason the complexity of phased planar arrays may not be warranted for hyperthermia applications although an array which simply offers amplitude control of individual elements appears promising, as discussed below. The performance of phased arrays in which the elements are positioned around the region of interest is more promising.

An analytical solution to focussed heating in cylindrical targets due to an array of N coherent horn antennas located around the circumference of the cylinder has been described by Wait (1985), Wait and Lumori (1986) and Lumori (1988). The dependence of the absorbed power distribution within a 12 cm diameter cylinder of muscle-like medium ($\varepsilon = 51 - j25$) was calculated using a Bessel function model for the cases of 4, 8 or 16 915-MHz aperture sources and it was shown that a local maximum in SAR could be achieved at the primary focus of the array although the global maximum in SAR was produced at the periphery of the cylinder. The 4-aperture array produced a primary focus which was less pronounced

than those of the 8- or 16-aperture arrays, secondary foci which were absent from the distributions produced by the larger arrays and relatively higher SAR near the surface of the cylinder. The existence of small gaps between the aperture sources around the circumference of the cylinder has been shown to have little or no effect on the performance of the array (Wait 1986).

A system consisting of eight tapered dipoles in the form of strip radiators located within a plastic cylinder and spaced from the tissue by a bolus has been described by Turner (1986). Four groups of two dipoles are symmetrically positioned around the cylinder and are usually fed in phase and with equal power by means of a four-way power divider. The array applicator is 31 cm long and has inner and outer diameters of 29 and 34 cm, respectively. The normal range of operating frequencies for this applicator is 105 – 180 MHz although an external tuner may be needed toward the extremes of this range and phantom studies have suggested that significant heating can be achieved on the central axis of cylindrical phantoms ($\varepsilon = 78 - j121$ at 120 MHz) up to 20 cm in diameter. Guerquin-Kern et al. (1987) measured SAR distributions produced by this device in a leg-shaped phantom ($\varepsilon = 62 - j88$ at 200 MHz) and found that the effective heating length was about 15 cm. The SAR distribution was relatively uniform within any cross-section of the phantom but exhibited significant variation along the axis of the leg, being higher in regions of smaller cross-section. Charny et al. (1986) measured SAR distributions in an amputated human leg. Their most significant finding was that the SAR in bone was significant, especially when the bone was positioned close to the axis of the array. SAR in muscle was highest in regions close to the axis and the ability to produce a hot spot in a desired region in the leg was limited. These authors concluded that relative phase and amplitude adjustments to the excitation of the dipoles would facilitate the ability to treat a wider number of locations in the limb. Coldefy et al. (1987) predicted and measured SAR distributions produced within a human leg by this array applicator for the usual case where the groups of dipoles were excited in phase and also for the case when four dipoles around the upper half of the applicator had a 180° phase shift with respect to the four dipoles around the lower half. For in phase excitation the maximum SAR was found in the central plane of the applicator and a fairly uniform distribution of SAR throughout deep regions of the leg's cross-section was found. The phase-shifted case produced maximum SAR in a plane 5 cm from the central plane of the applicator. A simulation of treatment of the lower leg using a whole body thermal model has been reported by Charny and Levin (1988).

In a theoretical study Jouvie et al. (1986) considered phase conjugation and amplitude inversion for both complete and limited annular arrays of small aperture sources at 434 or 1225 MHz around a realistic model of the neck. Focussing the complete array of ten 434-MHz apertures at a position in a posterior region resulted in a gain in power at the focus, albeit at the expense of higher SAR in tissues in the anterior regions. Several excitations of the limited annular array, which consisted of nine dielectrically loaded waveguides in a 3×3 configuration, were investigated and it was shown that phase control improved the SAR distribution within a targeted tumour volume compared with that due to a single 434-MHz applicator or a pair of 13.56-MHz capacitive electrodes. These authors commented that the SAR distribution was insensitive to changes of a few degrees in phase and 10% in amplitude.

Melek and Anderson (1981) used a ray-tracing technique to study the performance of an array of 17 dipoles radiating at 2450 MHz. The dipoles were arranged in semicircular or semi-elliptical arrays and an orthogonal short linear array and spaced by an air gap from an elliptic cylindrical phantom with complex permittivity $\varepsilon \approx 22 - j4.3$. Their calculations showed that a focus of the order of λ with a global maximum in SAR 3 dB greater than the maximum SAR at the surface could be produced at a depth of approximately one plane wave penetration depth (≈ 4.3 cm) for the phantom used. The phantom was intended to simulate lung tissue [$\varepsilon \approx 32 - j9.7$ at 2450 MHz (Iskander 1982)] and had a much lower loss (≈ 2 dB cm^{-1}) than muscle tissue (> 5 dB cm^{-1}). The relatively deep focus reported could not be expected in a patient due to the presence of the highly attenuating muscle layer which must be traversed before entering lung tissue.

Since attenuation in lung tissue is relatively low ($d \approx 6$ cm at 434 MHz) the question as to whether an array operating at a lower frequency might offer a feasible approach to hyperthermal treatment of lung tumours is being addressed. For example, Bach Andersen (1987) computed absorbed power distributions in a uniform lung–like medium ($\varepsilon \approx 25 - j18$) due to a phase and amplitude controlled array of four 400-MHz applicators placed at 90° intervals around the 20 cm $\times$ 20 cm region. An SAR of around 50% of the peak (which occurred beneath the applicators at the surface of the phantom) could be produced at a depth of 6 cm when the relative phases and amplitudes of the applicators were suitably adjusted. The effect of changing frequency in the range 300–

450 MHz was also investigated. At the low end of this range, the wavelength was too large to achieve a local maximum whilst at the high end of the range the shorter wavelength led to the formation of local maxima away from the primary focus, a feature which could be problematic if such arrays were to be used clinically.

Clearly, if multiple applicators are to be placed around a particular region of the body, the size of each applicator must be decreased if the number of applicators is increased. A trade-off must be made between the effective penetration from each applicator (favouring large applicators) and the gain due to constructive interference of the fields (favouring a larger number of applicators). The relationship between penetration and number of applicators has been discussed by Knoechel (1983) and Johnson et al. (1985). If the elements in an array are electrically small, any adequate bolus must be an integral part of the array to prevent excessive heating of superficial tissues.

1.4.5.3 Arrays with Large Effective Field Size

There is also interest in developing applicators capable of treating larger fields. Sandhu and Kolozsvary (1984) and Tanabe et al. (1983) have described microstrip applicators for this purpose. A prototype applicator capable of conforming to a cylindrical surface and intended for hyperthermal treatment of the upper arm has been described by Wilsey et al. (1988). The applicator consists of eight printed circuit plates, each containing two microstrip spiral antennas operating at 915 MHz, mounted on a water bolus. Adjacent pairs of antennas are excited noncoherently and power to each pair can be adjusted individually.

Hand et al. (1986) compared the effective field size defined by the -3 dB power level 2 cm into a muscle phantom for a 4×4 array of small aperture 434-MHz sources and a single aperture source with the same dimensions as the complete array (19 cm $\times$ 19 cm) and found that the effective area for the single aperture was 52% of that associated with the array. In an experimental study Hand et al. (1987) investigated a 4×2 array of distributed current applicators. An advantage of this type of applicator compared with a dielectrically loaded waveguide is that its effective field size is relatively large with respect to the dimensions of the applicator. The effective field size of the array was approximately 26 cm $\times$ 17 cm. A prototype applicator consisting of an array of eight parallel dipole antennas, isolated from each other by reflectors, which can be mechanically scanned within a water-filled foam matrix was reported by Lee et al. (1988). The treatment field for this device is 24 cm × 15 cm and the applicator is conformable to areas such as the chest wall.

Another approach to treating large surface areas is to employ mechanical scanning of the applicator. The compact nature and light weight of microstrip applicators are marked advantages for this approach. Kapp et al. (1988) describe two scanning applicators, one using a single spiral antenna, the other using two spiral antennas in which the antenna substrate is rotated 350° in a reciprocal movement by a stepper motor. The antenna is housed within a cylindrical structure and kept in contact with a membrane retaining a bolus.

1.4.6 Summary

1. Three types of applicator have been considered – capacitive or electric devices, inductive or magnetic devices and radiative devices.
2. Electric devices are relatively simple in construction but have the drawback that the electric field polarisation may lead to excessive heating of superficial fat tissue.
3. Induction coils produce an E-field in tissue which is parallel to skin-fat-muscle interfaces but the SAR distribution has a null at the central axis of the coil. More promising applicators are those employing a distributed current and which can be likened to an induction coil with its plane perpendicular to the tissue surface. These applicators may be considered in terms of magnetic dipole moments, with that of the coil applicator being perpendicular to the body's surface and that of the distributed current applicator being parallel to the body's surface.
4. Radiating applicators are often based on rectangular waveguides or microstrip. The TE_{10} mode is usually used for waveguide applicators. Ridged waveguides or differentially loaded waveguides were also considered.
5. The penetration depth of an applicator depends strongly on its dimensions in relation to the wavelength. Electrically small applicators have a penetration which is less than plane waves at the same frequency. Penetration also depends upon the curvature of tissues and the thickness of tissue layers relative to the wavelength.
6. Arrays of applicators can be used in either incoherent or coherent modes. Arrays designed to treat large areas without demanding significant penetration can be excited incoherently. An alternative approach is to scan an applicator over the region to be treated. If increased penetration and focussing are required, the array must be excited in a coherent manner. Quasi-planar arrays seem to offer little in terms of increased penetration but are useful in treating large areas. Arrays placed around curved tissue structures offer better localisation of SAR around target volumes and increased penetration. Phase conjugation with or without amplitude inversion provides a means of determining the excitation of array elements.

1.5 Applicators for Regional Hyperthermia

1.5.1 Electric (Capacitive) Applicators

1.5.1.1 Two-Electrode Systems

The deep heating characteristics of a capacitive applicator are strongly dependent upon the dimensions and separation of the electrodes. Although there have been frequent attempts in the past to use capacitive electrodes to induce deep body hyperthermia, the relatively small electrodes employed produced predominantly superficial heating as demonstrated by Armitage et al. (1983), who calculated SAR in anatomically real cross-sections. Kato et al. (1985) and Song et al. (1986) measured SAR in homogeneous muscle phantoms of different thicknesses due to electrodes with diameters ranging from 5 to 25 cm and demonstrated that if the diameters of the electrodes were equal to or greater than the thickness of the material heated, significant SAR could be produced at the centre of the phantom. A similar study, including the use of an inhomogeneous phantom which demonstrated the disturbance of current flow around a region of low conductivity simulating bone, was reported by Oleson and Cetas (1982). In Japan there has been much interest in developing radiofrequency capacitive hyperthermia systems using large electrodes with diameters of 20–25 cm operating at 8 or 13.56 MHz, and there is now widespread clinical use of the technique in that country (Koga 1988). The popularity of the technique has given incentive to a number of groups to develop two-dimensional thermal models based on patient CT scans (Ohguchi and Tsutsumi

1988; Kato et al. 1988). There appear to be two major limitations to the technique. Firstly, excessive SAR can be produced in superficial fat layers, and although skin cooling may reduce the problem in some cases (Matsuda et al. 1984), clinical experience suggests that the technique is not appropriate for patients with fat layers thicker than approximately 1.5–2 cm (Hiraoka et al. 1985). The second limitation is that current flow and current density in the electrically inhomogeneous body can lead to undesirable hot spots (Armitage et al. 1983; Kato et al. 1985; Oleson and Cetas 1982). With a two-electrode system, modification of the SAR distribution requires changes in shape, size and/or position of one or both electrodes.

Paulsen et al. (1988b) have made a theoretical study of the electric field distributions in a three-dimensional homogeneous model of man. A boundary element method was used to achieve adequate resolution (a nodal spacing of 5–8 cm) with moderate computing power. The capacitive applicator was modelled by a pair of parallel rectangular electrodes 35 cm (axial dimensions) × 32 cm (transverse dimension) separated by 34 cm and positioned over the lower abdominal region. Spacing from the body surface was typically 5–6 cm and the "background" medium was deionised water. In addition to large magnitudes associated with effects due to the edges of the electrodes, the electric field at the surface but within the body was highly dependent upon the local spacing between the irregular shaped tissue and the electrodes. Significant fields (10%–20% of the overall maximum) at the surfaces of the body's extremities were also predicted. Penetration into the body was relatively poor, with fields in large areas of the central regions of the body's cross-section being 20% or less of the maximum field at the surface. The authors pointed out that their study should be considered as a foundation for more detailed three-dimensional modelling in the future rather than representing electric field distributions produced in actual treatments.

1.5.1.2 Three-Electrode Systems

The inclusion of a third electrode in a capacitive system in principle offers the possibility of steering the SAR distribution by appropriate adjustments of the phases and amplitudes of the voltages applied to the three electrodes. Morand and Bolomey (1987a, b) have analysed such a system using a two-dimensional homogeneous rectangular model. The excitation of the electrodes can be considered in terms of a symmetrical common mode and an asymmetrical dif-

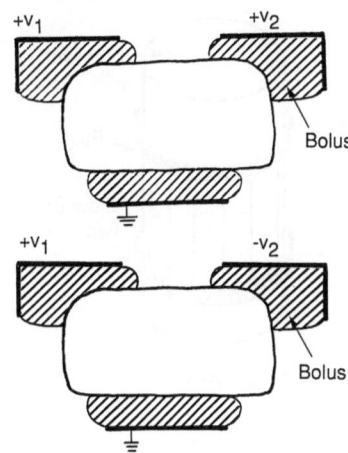

Fig. 1.29. Two methods of driving a three-electrode capacitive system: common mode (*upper diagram*), differential mode (*lower diagram*)

ferential mode (Fig. 1.29). By varying the magnitude or phase of the voltage at one of the upper electrodes with respect to the other these authors showed that a range of excitation conditions exists in which the current flow between the three electrodes and consequently power deposition can be modified significantly. However, the location of the highest power density remained near to the electrodes. In practice this problem would be accentuated due to the presence of superficial layers of fatty tissue which were not accounted for in the homogeneous model considered by Morand and Bolomey. Nussbaum et al. (1986) described experiments in which beef phantoms were heated by a three-electrode system. The power delivered to each of the electrodes could be varied and measurements within the phantom indicated that the vertical distribution of SAR could be modified significantly but that the lateral distribution at mid-depth of the phantom showed no appreciable change. Although high SAR in fat layers is an inherent problem for capacitive hyperthermia systems, the flexibility offered by three-electrode systems seems worthy of further investigation.

1.5.1.3 Ring Electrode Systems

There has also been interest in the use of ring-shaped capacitive electrodes placed around the body (Franconi 1987; van Rhoon et al. 1988), a configuration which can produce a circumferential axially directed electric field (Fig. 1.30). Von Rhoon et al. (1988) measured SAR distributions in circular cylindrical muscle phantoms due to rings 5 cm wide placed close to or

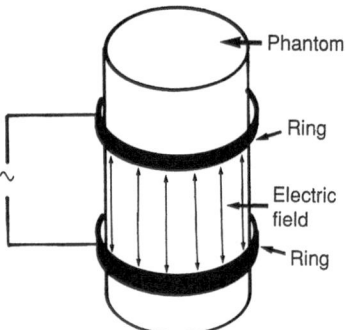

Fig. 1.30. Ring electrodes

directly on the plastic cover surrounding the phantoms. RF power at 27.12 or 13.56 MHz was applied between the rings whose separation was varied from 10 cm to 25 cm. At these low frequencies, the SAR distribution in the phantom is relatively uniform except in the region below the rings. Due to the field singularity at the edges of the rings, a gap between the ring and the phantom surface is necessary to avoid local hot spots below the rings.

1.5.2 Magnetic Applicators

1.5.2.1 Concentric Coil

A magnetic applicator which has been discussed at some length in the literature is the concentric coil (Storm et al. 1981), which consists of a single cylindrical electrode formed from copper sheet and placed around the patient concentrically with his or her longitudinal axis (Fig. 1.6). A 13.56-MHz current is driven through the electrode. A simple model which gives insight into this applicator is that of a circular cylinder of lossy medium placed concentrically with respect to a current loop. Neglecting the length of the actual applicator and assuming that the frequency is sufficiently low that there is negligible phase progression around the loop we have, using one of Maxwell's equations

$$\oint_1 \mathbf{E} \cdot \mathbf{dl} = -j\omega\mu_0 \int_S \mathbf{H} \cdot \mathbf{ds} \qquad (1.110)$$

The magnetic field is z-directed and the induced electric field is in the ϕ-direction (in cylindrical coordinates) and therefore parallel to skin-fat-muscle interfaces. E_ϕ is dependent upon radial distance r and is given by:

$$E_\phi(r) = -\frac{j\omega\mu_0 H_z r}{2} \qquad (1.111)$$

The SAR distribution is proportional to $(H_z r)^2$ and is zero at the centre of the load (r = 0). Armitage et al. (1983) and Hill et al. (1983) calculated SAR in two-dimensional realistic anatomical phantoms which were not necessarily concentrically located with respect to the applicator and showed that significant variations in SAR can occur over relatively short distances when tissues of low and high conductivity are adjacent, e.g. around the sternum and sacrum. A three-dimensional calculation of the induced electric field within a homogeneous man model (Paulsen et al. 1988 b) showed that the electric fields fall off rapidly at axial positions outside the applicator, which was taken to be 35 cm long, and suggested that the general results of the two-dimensional models of this device remain valid. Theoretical assessments of temperature distributions associated with a concentric coil applicator suggest that tissues within about 6 cm of the skin can be treated (Halac et al. 1983; Paulsen et al. 1984 a) and are in agreement with clinical experiences (Oleson et al. 1983 b, 1986; Sapozink et al. 1985).

1.5.2.2 Coaxial Coils

In another inductive technique (Fig. 1.31) a coaxial pair of coils is placed about the body to improve penetration over that associated with a single coil (Oleson 1984). Unlike the concentric coil, this arrangement can achieve significant SAR at positions off the common axis of the coils in coronal and sagittal planes in the patient. Oleson et al. (1983 a) suggest that the coil diameters should be within the range 18–28 cm and that the coils should be separated by approximately 30 cm. Smaller coils have inadequate penetration whilst larger coils can lead to excessive heating in the patient's flanks. Some improvement in

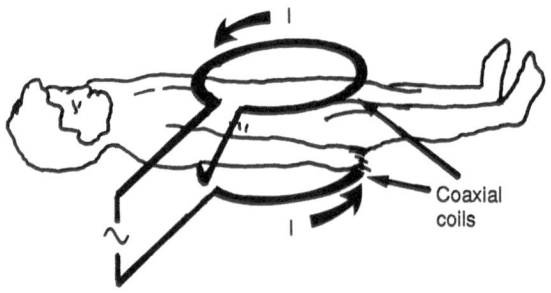

Fig. 1.31. Pair of coaxial coils

power deposition may be achieved by using solenoidal coils rather than planar ones. Von Ardenne et al. (1977) described a system in which the coils were scanned over the patient but Oleson (1984) reported experiments in which scanning increased the SAR near the boundary of a phantom rather than at depth. Oleson (1984) also drew attention to the need for modelling the change in SAR distribution within dielectrically inhomogeneous loads of complex shape due to movement of the coils. Corry et al. (1988) reported their clinical experience in using coaxial coils and found that small intrathoracic tumours could not be treated as well as tumours greater than 5 cm in diameter. Despite the null in SAR at the central axis of the coils, there was no tendency for lower temperatures to be induced in the central areas of tumours. Suboptimal heating in 28% of cases treated by this system were power limited (at 1000 W). The coaxial coil magnetic induction system performed very poorly in the thorax, where local pain was a severe limiting factor.

1.5.2.3 Helical Coil Applicators

Applicators based on a helical coil have the advantage that the axial component of current in the coil is associated with an axial induced electric field in the load which, for small diameter loads, is relatively uniform across planes perpendicular to the coil axis. Ruggera and Kantor (1984) reported self-resonant single layer helical coils suitable for heating bilayered arm or thigh phantoms. The lengths of the windings were adjusted to be either $\lambda/2$ or λ (half-wave or full-wave operation) and operating frequencies of 13.56, 27.12 or 40.68 MHz, which were dependent upon loading, were chosen. The pitch angles of these helical coils ranged from 1°–4°. The maximum of the axial electric field could be produced in the central regions of these coils and it was suggested that the uniformity of heating across transverse planes could be optimised when the coil length was twice the coil diameter for half-wave operation or four times the coil diameter for full-wave operation. Helical coil applicators designed to heat the trunk or thigh were investigated by Hagmann et al. (1985). The trunk helix was 43 cm long and 43 cm in diameter and the thigh helix was 36 cm long and 23 cm in diameter. These helical coil applicators had a length–diameter ratio of 1 and 1.57, respectively, which could be expected to limit the depth of heating (Ruggera and Kantor 1984). The authors noted that coils with a ratio closer to the ideal value could be difficult to use in treatments of some parts of the body.

The electric field distribution associated with a helix has axial, azimuthal and radial components and has been discussed by Vaughan and Bach Andersen (1985). Bach Andersen (1986) considered a sheath helix model, with dimensions similar to those of the helices described by Ruggera and Kantor (1984) (10 cm diameter, 3° pitch angle, 27 MHz), around a cylindrical muscle phantom 8 cm in diameter. It was shown that for these helices the E-field component which contributed the greatest to the absorbed power distribution was the axial E_z but for larger diameters (32 cm helix, 30 cm phantom) and the same pitch angle the azimuthal component E_ϕ was dominant. In this case the absorbed power distribution would resemble that associated with the concentric coil applicator. Increasing the pitch angle of the large helix to 25° improved the relative contribution from E_z. Hagmann (1984) and Hagmann and Levin (1984) analysed the performance of a sheath helix around a cylindrical layered arm phantom and suggested that a modest improvement in the ratio of power deposition in deep muscle to that in superficial fat could be achieved by reducing the operating frequency and increasing the pitch angle and/or radius of the helical winding. Since frequency and pitch angle are interdependent for self-resonant coils, Hagmann and Levin (1984) suggested that a better ratio might be achieved if the helix was operated at a lower frequency and tuned externally or if a travelling wave mode of operation was adopted. The coupling efficiency of helical coil applicators designed for heating the arm or thigh was investigated in phantoms by Hagmann and Levin (1985) and found to be in the range 56%–86%.

Hagmann (1988) reported a numerical analysis of the helical coil applicator in which the ratio of axial to surface heating in cylindrical muscle phantoms (radii 4–10 cm) was investigated. The results indicated that significantly larger pitch angles (>20°) and higher frequencies than those of earlier reported devices (1°–4° and 13.56–67 MHz) might be required. The extreme case of the pitch angle being equal to 90° (i.e. axially directed currents) produced the maximum ratio of axial to surface heating. This case corresponds to the axial electric fields described earlier by Brezovich et al. (1982) and Morita and Bach Andersen (1982) as being conducive for deep heating within lossy cylinders.

1.5.2.4 Other Magnetic Devices

An applicator consisting of a conducting sheet folded into a toroidal shape and supporting a distributed

current has been suggested by Franconi (1987). This
device is associated with an induced E-field which is
fairly constant in magnitude across the central region
of the torus at 27 MHz and directed predominantly
along the longitudinal axis of the torus.

Distributed current applicators operating at low RF
frequencies have also been suggested as the bases of
regional hyperthermia systems (Ikeda et al. 1988;
Johnson et al. 1987; Kato and Ishida 1983).

1.5.3 Radiative Applicators

1.5.3.1 Ridged Waveguides

Aperture type applicators have been used to produce
electric fields at frequencies in the range from about
120 to 27 MHz which are linearly polarised and directed predominantly along the patient's longitudinal
axis. This configuration is conducive to deep heating
in cylinders of lossy material (Brezovich et al. 1982;
Morita and Bach Andersen 1982). One approach has
been to use a pair of water-filled, single-ridged
waveguide applicators designed to operate at 27 MHz
such as those described by Paglione et al. (1981).
These applicators had effective apertures ranging
from approximately 25 cm × 8 cm to 30 cm × 14 cm
and included an integral water bolus to protect the
tissues from the high fields close to the aperture. According to van Rhoon et al. (1984) the effective
penetration depth in muscle phantom of these applicators is approximately 6 cm. SAR distributions
within the pelvic region due to a pair of applicators of
this type have been calculated by van den Berg et al.
(1983) and phantom experiments have been reported
by Visser et al. (1987).

1.5.3.2 Annular Arrays

An annular array device which consists of 16 parallel
plate waveguide apertures (20 cm × 23 cm) arranged in
pairs within an octagonal array has been developed by
Turner (1984 a, b). The length of the aperture is 46 cm
and the internal diameter of the array is 51 cm. In its
basic form, power from a 2-kW, 50 – 110-MHz amplifier is split by a four-way power divider to drive each
of four "quadrants" consisting of four apertures located in two adjacent faces of the octagonal array
with equal amplitudes and phases. If desired, individual quadrants may be turned off. The patient is located within the array and is exposed circumferentially to a cylindrically convergent E-field polarised along

the patient's longitudinal axis. The space between the
aperture sources and the patient is occupied by an integral water bolus which improves coupling to the patient, reduces stray field levels and controls skin temperature. Positioning the patient within the bolus and
array and the choice of operating frequency can be
aided by measuring the relative electric field strengths
at a few points at the skin – bolus interface. These
measurements give a qualitative indication of the
SAR distribution within the patient (Turner 1984a).
The longitudinally polarised E-field can give rise to
relatively high SAR in regions of the body remote
from the aperture of the array, notably in the neck
region and in the ankles and knees (Turner 1983; Hagmann and Levin 1986), although the problem can be
reduced by bolusing these regions with saline-filled
bags.

Iskander et al. (1982) used a moment method to calculate SAR in a two-dimensional realistic cross-section of the thorax due to the annular array operating
at 70 MHz and later (Iskander and Khoshdel-Milani
1984) calculated steady state temperature distributions for this case. The results were sensitive to the values allocated to blood perfusion rates and other thermal parameters in the model. Paulsen et al. (1984b,
1985) and Strohbehn et al. (1986) have calculated two-
dimensional distributions of (E-field)^2 and SAR
within heterogeneous tissues defined by CT scans of
patients. The scans included four pelvic tumours, two
abdominal tumours, two thoracic tumours and one
leg tumour. They showed that the distribution of (E-
field)^2 throughout the patient was such that a three-
or four-fold variation between skin and central region
locations could be expected and that the maximum
SAR tended to occur in the superficial muscle layer in
the hip region. SAR in tumour regions ranged from
20% to 60% of the maximum. Temperature distributions estimated using the bioheat transfer equation
suggested that large volumes of most of the tumours
failed to reach therapeutic temperatures, even when
low tumour blood perfusion rates were assumed.
These studies also suggested that this type of hyperthermia device might be more useful for treating
tumours in the pelvis rather than those in the thorax
or abdomen. Clinical experiences reported by Gibbs et
al. (1984), Gibbs and Stewart (1985), Howard et al.
(1986), Kapp et al. (1988), Oleson et al. (1986),
Sapozink et al. (1984, 1985, 1986, 1988) and Shimm
et al. (1988) indicate similar difficulties in reaching
therapeutic temperatures in deep tumours. Common
problems which have limited treatments include pain,
general patient discomfort and systemic heating.

A simulation of a hyperthermia treatment to the
pelvis using an annular array has been reported by

Charny and Levin (1988). In this work the absorbed power distribution was calculated using a three-dimensional block model of the body (Hagmann and Levin 1986) and used as input to a whole body thermal model (Charny et al. 1987). A substantial increase in cardiac output due mainly to vasodilation in superficial tissues and a moderate increase in oral temperature was predicted for this type of treatment. Paulsen et al. (1988b) modelled the electric field distribution in a three-dimensional homogeneous man model due to an annular array positioned over the lower abdomen. Significant field levels were predicted in the head and neck regions and below the knees. The field levels within cross-sections were relatively constant over large areas around the central axis and the effective length of the array (50% or more of the maximum electric field on the inside surface of the body) was approximately two-thirds of the axial dimension of the array.

The effects of adjusting relative phases and amplitudes of the fields from each aperture of the annular array have been investigated both theoretically and experimentally. Strohbehn et al. (1986) showed in a two-dimensional study that SAR within a tumour region could be improved considerably when phases and amplitudes were suitably adjusted. Sathiaseelan et al. (1986) demonstrated the feasibility of improving the temperature distribution within a patient by phase and amplitude adjustments. Samulski et al. (1987) compared the results of treatments in which all eight pairs of apertures received power with those in which only two adjacent quadrants close to the tumour were powered. A slight improvement in the temperatures monitored within tumour was observed during the two-quadrant treatments and this was achieved without increasing systemic stress or pain although higher normal tissue temperatures were observed. The authors concluded that any improvement resulting

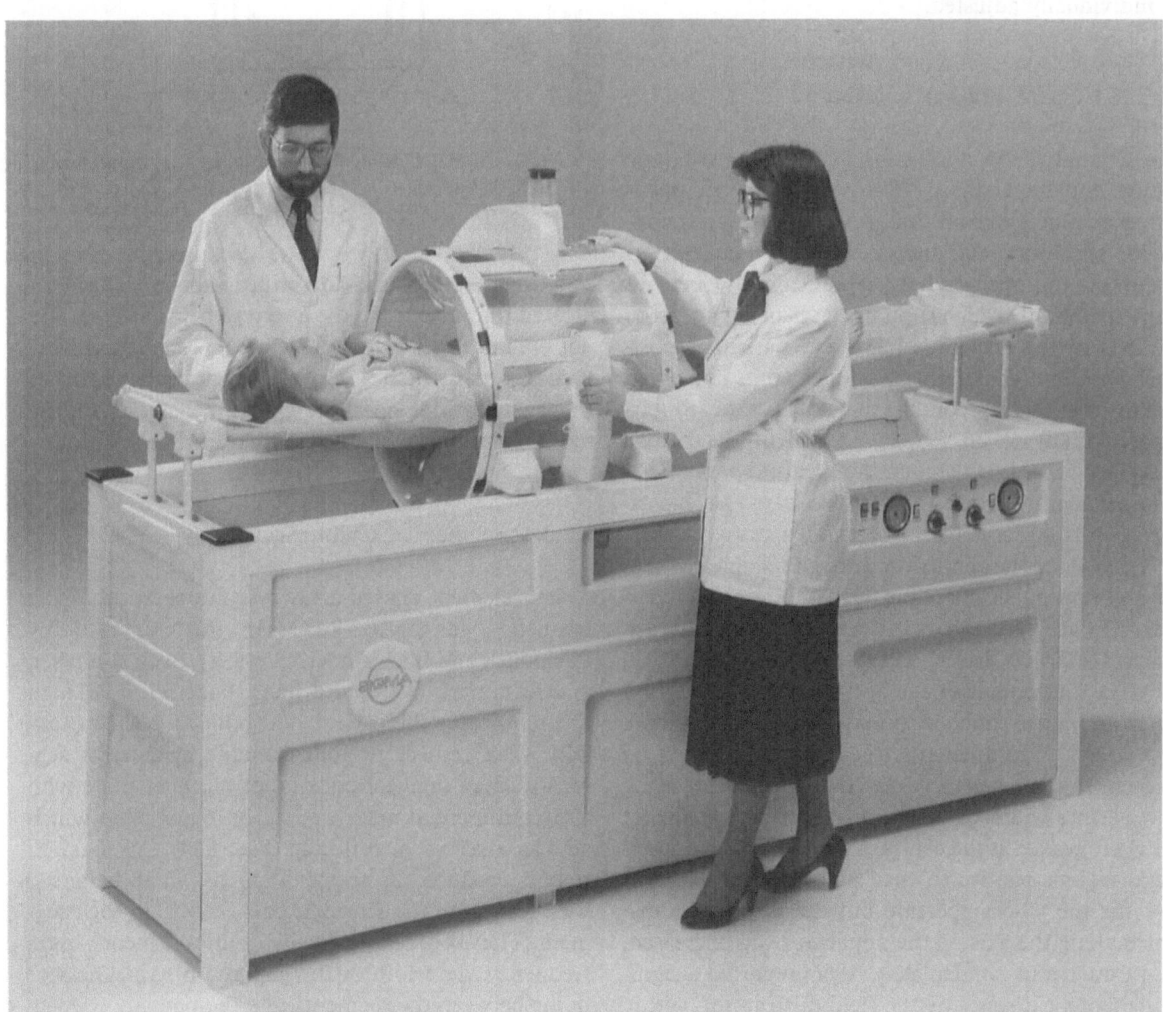

Fig. 1.32. Cylindrical array applicator (photography by courtesy of BSD Medical Corporation)

from a change in SAR distribution is likely to be small since tissue inhomogeneities and variations in blood flow appeared to be dominant factors in determining temperature distributions.

A new cylindrical array applicator designed for easier patient handling is described by Turner (1988). The array consists of eight dipoles dielectrically loaded by a water bolus and is contained within a transparent plastic cylinder (Fig. 1.32). Each dipole is approximately 46 cm in length and is aligned axially with the patient, who is positioned in the centre of the array. The internal diameter of the applicator is 58.5 cm and the bolus, which is contained by a silicone elastomer membrane, provides contact with the body for up to approximately 38 cm. The applicator operates at frequencies within the range 60–120 MHz and power is delivered to four pairs of adjacent dipoles at the top, bottom, left and right of the array. The relative phase and power associated with each of these groups may be individually adjusted.

1.5.3.3 Coaxial TEM Applicator

The coaxial TEM applicator (Lagendijk 1983; de Leeuw and Lagendijk (1987) consists of an open-ended air-filled coaxial line in which the inner conductor is hollow and has an internal diameter of 50 cm. At the open end the conductors are bent to form an aperture which supports an E-field parallel to the axis of the line (Fig. 1.33). The patient is positioned without constraint in a water bolus contained within the central conductor and is exposed circumferentially to an E-field parallel to the body's axis. This bolus arrangement should avoid patient discomfort arising from heavy bolus bags, as experienced with earlier versions of annular array devices. Power is fed to the applicator through a single connector and impedance matching to the feed line and generator is achieved through use of a tuner. The applicator may be driven at a frequency within 10–80 MHz. The length of the aperture formed between outer and inner conductors may be varied between 8 and 45 cm. Measurements in a cylindrical phantom 25.5 cm in diameter ($\varepsilon = 79 - j146$ at 70 MHz) using an aperture width of 20 cm or 40 cm indicated that the SAR peaked at the central axis. The SAR in superficial regions relative to that at the central axis was less for the 40-cm aperture but the half-power distance along the axis of the applicator was increased. Van Putten and van den Berg (1986) reported a rigorous theoretical study of the field distribution within the TEM applicator. The predictions of the effect of changing aperture width were in agreement with the

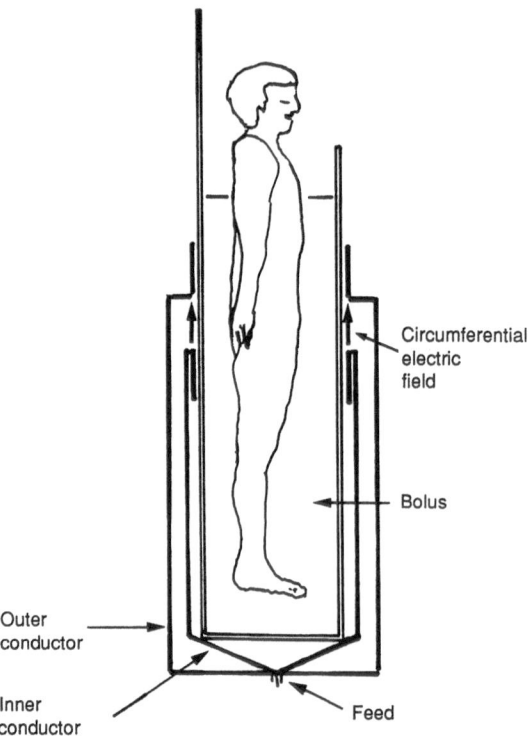

Fig. 1.33. Coaxial TEM applicator (after de Leeuw and Lagendijk (1987)

experimental findings described above. They indicated that increasing the aperture beyond 40 cm did not enhance the relative SAR at the central axis further and that a high SAR was present near the rim of the aperture due to the singular behaviour of the radial component of electric field at that position.

1.5.3.4 Segmented Cylindrical Array

A four-element phased array system operating over a frequency range 20–200 MHz was reported by Raskmark and Bach Andersen (1984), Bach Andersen and Raskmark (1985) and Bach Andersen (1987). In this applicator the patient plus bolus is placed coaxially with respect to four metallic cylindrical segments. Each segment consists of an inner plate with a circumferential aperture and an outer plate which acts as a screen or reflector (Fig. 1.34). An axial E-field is produced by applying an RF signal, balanced with respect to the screen, across the gap in each segment. The absorbed power distribution depends upon frequency, the width of the gap and the thickness of bolus between the segments and the patient. As with the coaxial TEM applicator, the fields are singular near the edges of the metallic segments and too nar-

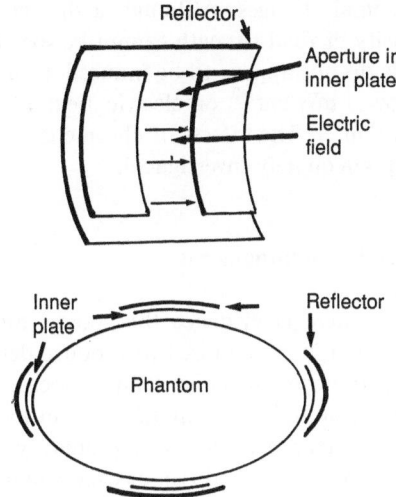

Fig. 1.34. Segmented cylindrical array. The *upper diagram* shows the aperture in the inner plate across which an electric field is maintained. Four such segments are placed around the body, as indicated in the *lower diagram*

row a gap (smaller than about a quarter of the wavelength in the lossy medium) leads to excessive absorbed power in superficial regions of the patient or load. The applicator has been designed with focussing through phase conjugation in mind (Knudsen and Hartmann 1986).

1.5.4 Summary

1. Three types of electric applicators for regional hyperthermia were considered. Those using two plate electrodes with a separation equal to or greater than their diameters are most commonly used, particularly in Japan. Limitations of the technique are excessive heating of superficial fat layers (especially if thicker than 1.5–2 cm, in which case cooling the skin is ineffective) and the inability to modify current flow within the electrically heterogeneous body. Newer applicators include systems with three electrodes which have been shown to give some control of SAR through phase and amplitude adjustments to the voltages at the electrodes and ring electrodes which produce an axially directed E-field circumferentially around the body.

2. Three types of magnetic applicator were discussed in detail. The concentric coil is useful for treatment of tumours within about 6 cm of the skin since a null in SAR is produced at its axis. A pair of coaxial coils can produce significant SAR deep within

the patient off his longitudinal axis but the distribution is highly non-uniform in the heterogeneous and geometrically irregular body. A marked difference between these two systems and the helical coil applicator is that the latter is associated with an axial electric field. Details of work reporting optimisation of helical coils were discussed and clinical evaluation is awaited.

3. The third approach to regional hyperthermia involves the use of radiative applicators. Most of the research effort to date has been aimed at developing and evaluating annular arrays or other circumferential apertures designed for use at frequencies between 27 and 120 MHz.

4. Clinical studies indicate that none of the electromagnetic regional hyperthermia devices investigated to date can consistently produce therapeutic temperatures throughout deep tumour volumes for significant periods of time. The development of applicators which are capable of delivering effective hyperthermia in deep tissues remains one of the major goals of hyperthermia research.

1.6 Biological Effects of RF/Microwave Fields and Exposure Standards

1.6.1 Biological Effects of RF/Microwave Fields

There are several thousand reports on biological effects of radiofrequency electromagnetic (RFEM) radiations in the scientific literature. Insight into this literature can be gained from recent books and symposia proceedings (e.g. Adair 1983; Baranski and Czerski 1976; Chou 1987; Hazzard 1977; Johnson and Shore 1976; Lin 1988; Marha et al. 1971; Osepchuk 1983; Presman 1970; Stuchly 1979), articles (e.g. Adey 1981; Carpenter 1977; Cleary 1983; Gandhi 1979; Michaelson 1982) and special issues of journals [e.g. *Annals of the New York Academy of Sciences* (1975) 247; *IEEE Engineering in Medicine and Biology* (1987) 6 (1); *IEEE Transactions Microwave Theory and Techniques* (1978) MTT-26 (8); *IEEE Transactions Microwave Theory and Techniques* (1984) MTT-32 (8); *Proceedings IEEE* (1980) 68 (1); *Radio Science* (1977) 12 (6S); *Radio Science* (1979) 14 (6S); *Radio Science* (1982) 17 (5S)]. In addition a digest of current literature is published periodically (e.g. Kleinstein 1987).

Some care is needed when interpreting the literature since many papers present unsubstantiated data or opinion. Furthermore, the complexities of the interaction between RFEM radiations and biological systems and the need for accurate dosimetry and temperature controls must be appreciated. This is particularly important when assessing much of the early data. Recently the reports of two review committees have been published (Elder and Cahill 1984; NCRP 1986) and these provide excellent overviews of this area of work. The following summary of biological effects of RFEM radiations consists of excerpts from the NCRP (1986).

1.6.1.1 Macromolecular and Cellular Effects

A large body of data has been collected on the effects of RF/microwave fields on biopolymers, particularly on enzymes, and in general the observed effects can be explained in terms of induced changes in temperature. Similarly, effects of fields on mitochondrial structure and function observed in a limited number of experiments seem related to temperature changes. There are reports in the literature describing studies on somatic cells exposed at high power densities. In many of these cases, however, it is not possible to exclude highly probable thermal effects because of the lack of sufficient detail regarding temperature control. Indeed, in more rigorous experiments effects were shown to be associated with temperature elevations. Results of some studies indicate that RF/microwave fields may produce metabolic effects on viral multiplication, but the lack of information on controls together with the finding that growth is very susceptible to changes in temperature must be borne in mind when considering this evidence. The evidence in the peer-reviewed literature is that athermal effects of microwave fields on cellular function (with the exception of effects on calcium efflux) are difficult to demonstrate. There are insufficient experimental data concerning cellular effects of microwaves at SARs below $1 \, W \, kg^{-1}$ for firm conclusions to be drawn.

1.6.1.2 Chromosomal Effects

Most investigations of chromosomal changes in mammalian cells have been carried out in vitro and at microwave frequencies. There is evidence that various types of chromosomal aberration may be induced and that high intensity fields induce chromosomal stickiness and breakdown. There is also a possibility that RF/microwave fields may induce specific chromosomal changes. Although a threshold for power density or field strength cannot be specified, such effects do not appear to be induced at power densities below $1 \, mW \, cm^{-2}$ or electric field strengths below $200 \, V \, m^{-1}$. The effects of chronic exposure have not been adequately investigated.

1.6.1.3 Carcinogenesis

Since there is evidence that power densities above $1 \, mW \, cm^{-2}$ are required to produce detectable levels of chromosomal damage and increase spontaneous rates of mutations, a tentative conclusion is that acute or short-term exposure to low intensity fields will not result in an increased risk of cancer in humans. However, it has been suggested that RF/microwave fields could be related, directly or indirectly, through physiological stress to carcinogenesis under certain (as yet undetermined) exposure conditions. The absence in the literature of well-documented evidence of an increase in the risk of cancer induction in humans or experimental animals arising from exposure to RF/microwave fields is inconclusive since only large effects could have been detected with the sample sizes and follow-up periods used.

1.6.1.4 Reproduction, Growth and Development

At high intensities, RF/microwave radiation can produce teratogenic effects. Whether these effects are primarily due to thermal stress or to some frequency-specific or field-specific action of the radiation, or to a combination of the two, has not been determined. In rodents, exposure during gestation to intense fields which are thermally stressful to the dam can result in a reduction in the weight of the fetus or an increase in incidence of resorptions. A few investigations report incidence of defects in the absence of measurable temperature change. The generality and implication of these results for human exposure have not been clearly demonstrated. Very few studies have been reported at frequencies that are highly penetrating in humans and at levels which appears to be non-thermal.

1.6.1.5 Haematopoietic and Immune Systems

Irradiation at SARs velow $1 \, W \, kg^{-1}$ and at frequencies from 300 kHz to 300 GHz, results in few, if any, unequivocal effects on the haematopoietic or immune systems of experimental animals. Under some exposure conditions, particularly for pulsed fields, the most consistent effects appear to be associated with an increase in the B-lymphocyte populations. There is a paucity of in vivo assays of immunocompetency associated with non-thermogenic exposures to RF/microwave fields. Even if future idealized experiments confirm the existence of effects on haematopoietic and immune systems in experimental animals, the extrapolation of these results to man and prediction of potential adverse effects will be difficult.

1.6.1.6 Endocrine System

The effects of exposure to RF electromagnetic fields on endocrine function are in general consistent with both immediate and long-term responses to thermal input and to non-specific stress, which can also arise from thermal loading. Available information on effects of RFEM exposure is consistent with neuroendocrine involvement in the many physiological adjustments of the organism relative to increased body temperature or to changes in temperature gradients within the body that could affect any individual component or combination of components in the neuroendocrine system. Evidence based on results of studies in rats suggests that endocrine disturbance in humans should not occur below an average SAR of $0.4\,W\,kg^{-1}$.

1.6.1.7 Cardiovascular Function

Whole body or regional exposure to RFEM levels which produce significant heating causes cardiovascular reactions similar to those associated with conventional heating of the body. Short-term exposure to low-level fields (below $10\,mW\,cm^{-2}$ or $2\,W\,kg^{-1}$) does not appear to produce cardiovascular effects. The few data available on long-term exposure are contradictory.

1.6.1.8 Blood-Brain Barrier

Although early reports indicated that exposure to RFEM radiation could result in increased permeability of the blood-brain barrier, later reports indicated that increased permeability might be mediated by hyperthermia and that cerebral circulation and not barrier function is influenced by weaker RFEM fields. There is histological evidence that moderately strong RFEM fields doe not impair the integrity of the tight junctions of the blood-brain barrier. There is no

evidence to suggest that any changes following brief exposures to weak-to-moderate fields result in irreversible neurological or behavioural disorders in experimental animals. The effects of truly long-term irradiation have not been investigated.

1.6.1.9 Nervous System

Effects of RFEM fields attributable to elevated temperature have been well categorized and generally are similar to effects of hyperthermia induced by other means. However, there are some interactions which induce exceedingly small changes in temperature ($\ll 0.1\,^{\circ}C$) yet which result in major physiological changes. In some interactions in brain and in some other tissues very weak fields within certain frequency and amplitude "windows" can produce a variety of behavioural, physiological and chemical responses by what appears to be amplification involving long-range, resonant molecular interactions. Other interactions with brain tissue are associated with amplitude-modulated RFEM fields for which the modulation frequency is less than 1 kHz but particularly in the range 1–20 Hz.

1.6.1.10 Cataractogenesis

Laboratory studies in rabbits have clearly demonstrated that absorption of RFEM energy can damage ocular tissues. The location of damage is dependent upon the frequency of the radiation and the mode of exposure whilst the degree of damage primarily depends upon the power density of the field, the absorbed energy and the duration of exposure (Cleary 1970). For frequencies in the range 1–10 GHz, the lens is the most sensitive tissue. Cataracts develop following either a single, high threshold dose or repeated subthreshold doses delivered in close succession. The limited body of data on cataracts induced by acute, high intensity RFEM exposure of humans indicates the involvement of thermal damage to lens tissue. No evidence of deleterious effect was indicated by the results of epidemiological studies of the relationship between occupational RFEM exposure and ocular changes.

1.6.2 Exposure Guidelines

Although guidelines for limiting exposure to RFEM fields have been formulated in North America and in many Western and Eastern European countries, there is no general consensus on acceptable limits. From 1966 to 1982 the American National Standards Institute (ANSI) standard C95.1 recommended a plane wave free-space power density limit of $10 \, \text{mW cm}^{-2}$. This limit was based on arguments of heat stress in man and thermal damage to tissue. The guidelines in many Western European countries were similar. In Eastern European countries and in the USSR, guidelines recommended levels which were significantly lower ($0.1 \, \text{mW cm}^{-2}$ for continuous occupational exposure, USSR 1976). These differences have been discussed by several authors (e.g. Dodge and Glaser 1977; Sliney et al. 1985).

Recently revisions (or proposals for revisions) of guidelines have been made which take into account current knowledge of whole and partial body resonances and internal dosimetry. These new guidelines cover wider frequency ranges and include frequency dependencies. In general there has been a downward revision from the $10 \, \text{mW cm}^{-2}$ limit in Western guidelines and an upward revision in Soviet standards (USSR 1984). A comparison of Polish, Czechoslovakian and Russian standards has been presented by Czerski (1985). Several agencies [e.g. Commission of the European Communities (CEC) 1980, see also Seguin 1983; ANSI 1982; American

Conference of Governmental Industrial Hygienists (ACGIH) 1983; National Radiological Protection Board (NRPB) 1986; International Radiation Protection Association (IRPA) 1984] recommend that occupational exposure should not exceed an SAR of $0.4 \, \text{W kg}^{-1}$ averaged over the whole body or a spatial peak value of $4 \, \text{W kg}^{-1}$ [$8 \, \text{W kg}^{-1}$ for ANSI (1982)] when averaged over any 1 g of tissue. In all cases the SAR is time averaged over any 0.1-h period. In some cases, at frequencies typically lower than a few megahertz, the limit is defined in terms of the electric or magnetic field strengths. Details of the asymptotic equations for RFEM exposure guidelines recommended by US and European agencies are given in Tables 1.13 and 1.14, respectively. The original references should be consulted for details of conditions and exclusions associated with particular guidelines.

1.6.3 Safety Procedures for Electromagnetic Hyperthermia

It is certainly prudent to meet an appropriate RFEM exposure guideline when designing or using RF/microwave hyperthermia equipment. For this reason hyperthermia devices and treatment areas should be

Table 1.13. Asymptotic equations for RFEM exposure guidelines (United States)

Frequency f (MHz)	(Electric field)^2 (V^2 m^{-2})	(Magnetic field)^2 (A^2 m^{-2})	Equivalent power density (mW cm^{-2})
ANSI (1982)			
0.3 – 3	400 000	2.5	100
3 – 30	4000·900/f^2	0.025·900/f^2	900/f^2
30 – 300	4000	0.025	1.0
300 – 1500	4000·f/300	0.025·f/300	f/300
1500 – 10^5	20 000	0.125	5.0
ACGIH (1983)			
0.01 – 3	377 000	2.65	100
3 – 30	3770·900/f^2	0.027·900/f^2	900/f^2
30 – 100	3770	0.027	1.0
100 – 1000	3770·f/100	0.027·f/100	f/100
1000 – 3×10^5	37 700	0.265	10.0

Frequency f (MHz)	Electric field (RMS) (V m^{-1})	Magnetic field (RMS) (A m^{-1})	Equivalent power density (mW cm^{-2})
IRPA (1984): Occupational exposure limits			
0.1 – 1	194	0.51	–
1 – 10	194/f$^{0.5}$	0.51/f$^{0.5}$	–
10 – 400	61	0.16	1
400 – 2000	3·f$^{0.5}$	0.008·f$^{0.5}$	f/400
2000 – 3×10^5	137	0.36	5

surveyed and any zones in which field levels exceed the chosen guideline should be clearly indicated.

Stuchly et al. (1983) reported measurements of RF fields around three concentric coil applicators (whole body coil, neck coil and thigh coil) driven at 13.56 MHz and loaded by either a human manikin filled with saline solution ($\sigma = 5.3$ mS m^{-1}) or, in a few cases, by human patients. The phantom provided a good simulation of the loading of the coil by a patient. Minimum distances from the coils (driven at 1 kW) for compliance with exposure to several levels of equivalent power density were determined. For example, at the 4.9 mW cm^{-2} (900/f^2 – see Table 1.13) level, minimum distances ranged from 0.61 m (perpendicular to coil axis) to 1 m (along coil axis) for the body coil and from 0.46 m (perpendicular to coil axis) to 0.85 m (along coils axis) for the neck coil. At the 1 mW cm^{-2} level, the corresponding distances were 0.95 and 2.0 m (for the body coil) and 0.75 and 1.25 m (for the neck coil).

In another survey Tofani and Agnesod (1984) measured stray radiation around RF (27.12 MHz) capacitive applicators 11 cm in diameter placed around either rectangular parallelepiped or cylindrical phantoms filled with saline solution (0.9% NaCl). A variety of electrode/phantom configurations were studied and a theoretical model for predicting electric field levels as a function of exposure geometry was presented. Exposure of operators of short-wave diathermy devices may exceed 60 V m^{-1} and/or 0.16 A m^{-1} (Stuchly et al. 1982). Ruggera (1980) also reported

measurements of emission levels during microwave (2450 MHz) and short-wave (27 MHz) diathermy treatments. Stray radiation around RF devices is generally higher than that for microwave devices. Leakage fields from microwave applicators depend upon the contact area between tissue and applicator and upon the orientation of the E-field vector with respect to the tissues (Lehmann et al. 1979). The possibility of other hazards associated with the operation of electromagnetic hyperthermia systems in a clinical environment should also be recognised (Stauffer and Hevezi 1982).

It is important that hyperthermia systems meet the electrical safety requirements demanded of other medical equipment, e.g. complicance with IEC regulation 601 or other international safety codes.

1.6.4 Summary

1. A summary of the biological effects of radiofrequency electromagnetic radiations according to the NCRP Report No. 86 was presented.
2. A summary of occupational exposure guidelines formulated by several US and European agencies was presented.
3. The need for surveys of stray field levels associated with hyperthermia devices was highlighted.
4. The need for electrical safety of hyperthermia devices and care in their use in a clinical environment was stated.

Table 1.14. Asymptotic equations for RFEM exposure guidelines (Europe)

Frequency f (MHz)	Electric field (RMS) (V m^{-1})	Magnetic field (RMS) (A m^{-1})	Equivalent power density (W m^{-2})
NRPB (1986): Whole-body occupational exposure (<2 h per day)			
0.3 – 10	600/f	5.0/f	–
10 – 30	60	5.0/f	–
30 – 100	60	0.16	10
100 – 500	$6 \cdot f^{0.5}$	$0.016 \cdot f^{0.5}$	f/10
500 – 3×10^5	135	0.36	50
DIN 57 848 Teil 2 (1984)			
0.3 – 2	1500	7.5/f	–
2 – 30	3000/f	7.5/f	–
30 – 3000	100	0.25	25
3000 – 12 000	$100 \cdot (f/3000)^{0.5}$	$0.25 \cdot (f/3000)^{0.5}$	$25 \cdot f/3000$
12 000 – 3×10^6	200	0.5	100
Italy: Normal exposure of workers (Seguin 1983)			
0.3 – 3	140	0.36	50
3 – 3×10^5	60	0.17	10

References

ACGIH (1983) American Conference of Governmental Industrial Hygienists. Documentation for threshold limit values (TLV's) for chemical substances and physical agents in the work environment with intended changes. ACGIH, Cincinnati

Adair ER (ed) (1983) Microwaves and thermoregulation. Academic, New York

Adey WR (1981) Tissue interactions with nonionizing electromagnetic fields. Physiol Rev 61:435–514

Allen S, Kantor G, Bassen H, Ruggera P (1988) CDRH RF phantom for hyperthermia systems evaluation. Int J Hyperthermia 4:17–23

Andreuccetti D, Bini M, Ignesti A, Olmi R, Rubino N, Vanni R (1988) Use of polyacrylamide as a tissue equivalent material in the microwave range. IEEE Trans Biomed Eng BME 35:275–277

ANSI (1982) American national standard safety levels with respect to human exposure to radiofrequency electromagnetic fields, 300 kHz to 100 GHz. ANSI C95.1-1982. IEEE, New York

Antolini R, Cerri G, DeLeo R (1984) Influence of the bolus on the radiation characteristics of waveguide applicators. In: Overgaard J (ed) Hyperthermic oncology, vol 1. Taylor and Francis, London, pp 651–654

Armitage DW, LeVeen HH, Pethig R (1983) Radiofrequency induced hyperthermia: computer simulation of specific absorption rate distributions using realistic anatomical models. Phys Med Biol 28:31–42

Audet J, Bolomey JC, Pichot C, N'Guyen DD, Chive M, Leroy Y (1980) Electrical characteristics of waveguide applicators for medical applications. J Microwave Power 15:177–186

Bach Andersen J (1985) Theoretical limitations on radiation into muscle tissue. Int J Hyperthermia 1:45–56

Bach Andersen J (1986) Regional electromagnetic heating. In: Hand JW, James JR (eds) Physical techniques in clinical hyperthermia. Research Studies, Letchworth, pp 65–97

Bach Andersen J (1987) Electromagnetic power deposition: inhomogeneous media, applicators and phased arrays. In: Field SB, Franconi C (eds) Physics and technology of hyperthermia. Nijhoff, Dordrecht, pp 159–188

Bach Andersen J, Raskmark P (1985) A regional hyperthermia phases array system. Proceedings 7th annual conference of the Engineering in Medicine and Biology society. IEEE, New York, pp 331–333

Bach Andersen J, Baun A, Harmark K, Heinzl L, Raskmark P, Overgaard J (1984) A hyperthermia system using a new type of inductive applicator. IEEE Trans Biomed Eng BME 31:21–27

Bahl IJ, Stuchly SS (1980) Analysis of a microstrip covered with a lossy dielectric. IEEE Trans Microwave Theory Tech MTT 28:104–109

Bahl IJ, Bhartia P, Stuchly SS (1982) Design of microstrip antennas covered with a dielectric layer. IEEE Trans Antennas Propag AP 30:314–318

Bahl IJ, Stuchly SS, Stuchly MA (1980) A new microstrip radiator for medical applications. IEEE Trans Microwave Theory Tech MTT 28:1464–1468

Balzano Q, Garay O, Steel FR (1979) An attempt to evaluate the exposure of operators of portable radios at 30 MHz. In: Proceedings of the 29th IEEE Vehicular Technology Society conference, Arlington Heights. IEEE, New York, pp 187–189

Baranski S, Czerski P (eds) (1976) Biological effects of microwaves. Dowden, Hutchinson and Ross, Stroudsburg

Bardati F (1986) Models of electromagnetic heating and radiometric microwave sensing. In: Hand JW, James JR (eds) Physical techniques in clinical hyperthermia. Research Studies, Letchworth, pp 327–382

Bassen HI, Coakley RF (1981) United States radiation safety and regulatory considerations for radiofrequency hyperthermia systems. J Microwave Power 16:215–216

Bassen HI, Smith GS (1983) Electric field probes – a review. IEEE Trans Antennas Propag AP-31:710–718

Batchman TE, Gimpelson G (1983) An implantable electric-field probe of submillimeter dimensions. IEEE Trans Microwave Theory Tech MTT 31:745–751

Beyne L, de Zutter D (1988) Power deposition of a microstrip applicator radiating into a layered biological structure. IEEE Trans Microwave Theory Tech MTT 36:126–131

Binger CAL, Christie RV (1927) An experimental study of diathermy: the conditions necessary for the production of local heat in the lungs. J Exp Med 46:585–594

Bini MG, Ignesti A, Millanta L, Olmi R, Rubino N, Vanni R (1984) The polyacrylamide as a phantom material for electromagnetic hyperthermia studies. IEEE Trans Biomed Eng BME 31:317–322

Bini M, Ignesti A, Millanta L, Olmi R, Rubino N, Vanni R (1985) An unbalanced electric applicator for RF hyperthermia. IEEE Trans Biomed Eng BME 32:638–642

Boeckel G (1873) Galvanocaustique thermique. Paris

Borup DT, Gandhi OP (1984) Fast-Fourier-transform method for calculation of SAR in finely discretized inhomogeneous models of biological bodies. IEEE Trans Microwave Theory Tech MTT 32:730–746

Borup DT, Gandhi OP (1985) Calculation of high-resolution SAR distributions in biological bodies using the FFT algorithm and conjugate gradient method. IEEE Trans Microwave Theory Tech MTT 33:417–419

Bozzetti M, DeLeo T, Ercoli C (1983) Energy absorption from waveguides in biological-like media. Alta Freq 52:185–187

Brezovich IA, Lilly MB, Durant JR, Richards DB (1981) A practical system for clinical radiofrequency hyperthermia. Int J Radiat Oncol Biol Phys 7:423–430

Brezovich IA, Young JH, Atkinson WJ, Wang MT (1982) Hyperthermia considerations for a conducting cylinder heated by an oscillating electric field applied parallel to the cylinder axis. Med Phys 9:746–748

Carpenter RL (1977) Microwave radiation. In: Lee DHK, Falk HL, Murphy SD (eds) Handbook of physiology no 9. Reactions to environmental agents. American Physiology Society, Bethesda, pp 111–125

Carpenter CM, Page AB (1930) Production of fever in man by short radio waves. Science 71:450–452

CEC (1980) Commission of the European Communities proposal for microwave and radiowave exposure standard. CEC, Luxembourg

Cetas TC (1982) The philosophy and use of tissue-equivalent electromagnetic phantoms. In: Nussbaum GH (ed) Physical aspects of hyperthermia. American Institute of Physics, New York, pp 441–461

Charny CK, Levin RL (1988) Simulations of MAPA and APA heating using a whole body thermal model. IEEE Trans Biomed Eng BME 35:362–371

Charny CK, Guerquin-Kern JL, Hagmann MJ, Levin SW, Lack EE, Sindelar WF, Zabell A, Glatstein E, Levin RL (1986) Human leg heating using a mini-annular phased array. Med Phys 13:449–456

Charny CK, Hagmann MJ, Levin RL (1987) A whole body thermal model of man during hyperthermia. IEEE Trans Biomed Eng BME 34:375–387

Cheever E, Leonard JB, Foster KR (1987) Depth of penetration of fields from rectangular apertures into lossy media. IEEE Trans Microwave Theory Tech MTT 35:865–867

Chen KM, Guru BS (1977) Internal EM field and absorbed power density in human torsos induced by 1–500 MHz EM waves. IEEE Trans Microwave Theory Tech MTT 25: 746–755

Chen TS (1957) Calculation of parameters of ridge waveguides. IRE Trans Microwave Theory Tech MTT 5:12–17

Cheung AY, Dao T, Robinson JE (1977) Dual-beam TEM applicator for direct contact heating of dielectrically encapsulated malignant mouse tumor. Radio Sci 12 (6S):81–85

Chou CK (1987) Biological effects of electromagnetic waves. In: Field SB, Franconi C (eds) Physics and technology of hyperthermia. Nijhoff, Dordrecht, pp 319–353

Chou CK, Chen GW, Guy AW, Luk KH (1984) Formulas for preparing phantom muscle tissue at various radiofrequencies. Bioelectromagnetics 5:435–441

Christie RV, Loomis AL (1929) The relationship of frequency to physiological effects of high frequency currents. J Exp Med 49:303–321

Christie RV, Ehrlich W, Binger CAL (1928) An experimental study of diathermy: the elevation of temperature in the pneumonic lung. J Exp Med 47:741–755

Cleary SF (1970) Biological effects of microwave and radiofrequency radiation. Crit Rev Environ Control 1:257–306

Cleary SF (1983) Bioeffects of microwave and radiofrequency radiation. In: Storm FK (ed) Hyperthermia in cancer therapy. Hall Medical, Boston, pp 545–566

Coldefy HM, Charny CK, Levin RL (1987) Theoretical and experimental results of power deposition in human legs irradiated by a MAPA. In: Proceedings of 9th annual conference of IEEE Engineering in Medicine and Biology Society, Boston, Nov 1987, vol 4. IEEE, New York, pp 1949–1950

Collin RE (1960) Field theory of guided waves. McGraw-Hill, New York, chap 7

Corry PM, Jabboury K, Kong JS, Armour EP, McGraw FJ, LeDuc T (1988) Evaluation of equipment for hyperthermia treatment of cancer. Int J Hyperthermia 4:53–74

Council on Physical Therapy (1934) Hyperpyrexia produced by physical agents. JAMA 103:1308–1309

Czerski P (1985) Radiofrequency radiation exposure limits in Eastern Europe. J Microwave Power 20:233–239

de Leeuw AAC, Lagendijk JJW (1987) Design of a deep-body hyperthermia system based on the 'coaxial TEM' applicator. Int J Hyperthermia 3:413–421

Denier A (1936) Les ondes herziennes ultracourtes de 80 cm. J Radio Electrol 20:193–197

DIN (1984) Hazards by electromagnetic fields. Protection of persons in the frequency range 10 kHz to 3000 GHz. DIN 57848, Part 2. Deutsche Elektrotechnische Kommission im DIN und VDE (DKE), Berlin

Dodge CH, Glaser ZR (1977) Trends in nonionizing electromagnetic radiation bioeffects research and related occupational health aspects. J Microwave Power 12:319–334

Doss JD (1982) Calculations of electric fields in conductive media. Med Phys 9:566–573

Durney CH (1980) Electromagnetic dosimetry for models of humans and animals: a review of theoretical and numerical techniques. Proc IEEE 68:33–40

Durney CH (1987) Calculation of electromagnetic power deposition. In: Field SB, Franconi C (eds) Physics and technology of hyperthermia. Nijhoff, Dordrecht, pp 152–158

Edenhofer P (1983) Field characteristics of a dual antenna sensor system probing biological tissues. Proceedings URSI symposium on electromagnetic theory, Santiago de Compostela, Spain, Aug 1983, pp 685–688

Eidinow A (1935) Discussion on short wave diathermy. Proc R Soc Med 28:312–315

Elder JA, Cahill DF (1984) Biological effects of radiofrequency radiation. Report EPA-600/8-83-026F. Health Effects Research Laboratory, Office of Research and Development, US EPA, Research Triangle Park, NC

Franconi C (1987) Hyperthermia heating technology and devices. In: Field SB, Franconi C (eds) Physics and technology of hyperthermia. Nijhoff, Dordrecht, pp 80–122

Franconi C, Tiberio CA, Raganella L, Begnozzi L (1986) Low frequency RF twin-dipole applicator for intermediate depth hyperthermia. IEEE Trans Microwave Theory Tech MTT 34:612–619

Fray C, Khayata N, Papiernik A (1982) TM_{10} admittance and radiation from a flanged open-ended waveguide in layered absorbing media. Arch Elektron Übertragungstech 36:107–110

Furukawa M, Kato H, Fujita Y, Uchida N, Kasai T, Ishida T (1988) Clinical trials of hyperthermia with inductive aperture-type applicator. In: Koga S (ed) Hyperthermic oncology in Japan '87. Imai, Yonago, Japan, pp 101–102

Gajda G, Stuchly MA, Stuchly SS (1979) Mapping of the near field pattern in simulated biological tissues. Electron Lett 15:120–121

Gandhi OP (1979) Dosimetry – the absorption properties of man and experimental animals. Bull NY Acad Med 55:990–1020

Gandhi OP, DeFord JF, Kanai H (1984) Impedance method for calculation of power deposition patterns in magnetically induced hyperthermia. IEEE Trans Biomed Eng BME 31:644–651

Gee W, Lee S-W, Bong NK, Cain CA, Mittra R, Magin RL (1984) Focused array hyperthermia applicator: theory and experiment. IEEE Trans Biomed Eng BME 31:38–46

Gibbs FA, Stewart JR (1985) Regional hyperthermia in the treatment of cancer: a review. Cancer Invest 3:445–452

Gibbs FA, Sapozink MD, Gates KS, Stewart JR (1984) Regional hyperthermia with an annular phased array in the experimental treatment of cancer: report of work in progress with a technical emphasis. IEEE Trans Biomed Eng BME 31:115–119

Gosset A, Gutmann A, Lakhovsky G, Magrou J (1924) Essais de therapeutique du 'cancer experimental des plantes'. C R Soc Biol (Paris) 91:626–628

Guerquin-Kern JL, Hagmann MJ, Levin RL (1987) Experimental characterization of the miniannular phased array as a hyperthermia applicator. Med Phys 14:674–680

Guo TC, Guo WW, Larsen LE (1984) A local field study of a water-immersed microwave antenna array for medical imagery and therapy. IEEE Trans Microwave Theory Tech MTT 32:844–854

Guy AW (1971a) Analysis of electromagnetic fields induced in biological tissues by thermographic studies on equivalent phantom models. IEEE Trans Microwave Theory Tech MTT 19:205–214

Guy AW (1971b) Electromagnetic fields and relative heating patterns due to a rectangular aperture source in direct contact with bilayered biological tissue. IEEE Trans Microwave Theory Tech MTT 19:214–223

Guy AW (1984) History of biological and medical applications of microwave energy. IEEE Trans Microwave Theory Tech MTT 32:1182–1200

Guy AW, Lehmann JF (1966) On the determination of an optimum microwave diathermy frequency for a direct contact applicator. IEEE Trans Biomed Eng BME 13:76–87

Guy AW, Lehmann JF, Stonebridge JB (1974) Therapeutic applications of electromagnetic power. Proc IEEE 62: 55–75

Hagmann MJ (1984) Propagation on a sheath helix in a coaxially layered lossy dielectric medium. IEEE Trans Microwave Theory Tech MTT 32:122–126

Hagmann MJ (1988) Optimization of helical coil applicators for hyperthermia. IEEE Trans Microwave Theory Tech MTT 36:148–150

Hagmann MJ, Levin RL (1984) Analysis of the helix as an RF applicator for hyperthermia. Electron Lett 20:337–338

Hagmann MJ, Levin RL (1985) Coupling efficiency of helical coil hyperthermia applications. IEEE Trans Biomed Eng BME 32:539–540

Hagmann MJ, Levin RL (1986) Aberrant heating: a problem in regional hyperthermia. IEEE Trans Biomed Eng BME 33:405–411

Hagmann MJ, Levin RL, Turner PF (1985) A comparison of the annular phased array to helical coil applicators for limb and torso hyperthermia. IEEE Trans Biomed Eng BME 32:916–927

Halac S, Roemer RB, Oleson JR, Cetas TC (1983) Magnetic induction heating of tissue: numerical evaluation of tumor temperature distributions. Int J Radiat Oncol Biol Phys 9:881–981

Hand JW (1987) Electromagnetic applicators for non-invasive hyperthermia. In: Field SB, Franconi C (eds) Physics and technology of hyperthermia. Nijhoff, Dordrecht, pp 189–210

Hand JW, Cheetham JL, Hind AJ (1986) Absorbed power distributions from coherent microwave arrays for localized hyperthermia. IEEE Trans Microwave Theory Tech MTT 34:484–489

Hand JW, Hind AJ (1986) A review of microwave and rf applicators for localised hyperthermia. In: Hand JW, James JR (eds) Physical techniques in clinical hyperthermia. Research Studies, Letchworth, pp 98–148

Hand JW, Hind AJ, Cheetham JL (1985) Multielement microwave array applicators for localized hyperthermia (Abstract). Strahlentherapie 161:535

Hand JW, Johnson RH (1986) Field penetration from electromagnetic applicators for localized hyperthermia. In: Bruggmoser G, Hinkelbein W, Engelhardt R, Wannemacher M (eds) Locoregional high-frequency hyperthermia and temperature measurement. Springer, Berlin Heidelberg New York, pp 7–17 (Recent results in cancer research, vol 101)

Hand JW, Ledda JL, Evans NTS (1982) Considerations of radiofrequency induction heating for localised hyperthermia. Phys Med Biol 27:1–16

Hand JW, Johnson RH, James JR (1987) A microwave hyperthermia system with multi-element applicator for treatment of superficial tumours. Int J Hyperthermia 3:566–567

Hand JW, Paulsen KD, Lumori MLD, Gopal MK, Cetas TC, Alkhairi S (1989a) Microwave array applicators for superficial hyperthermia. In: Sugahara T, Saito M (eds) Hyperthermic oncology 1988, vol 1, Taylor and Francis, London (in press)

Hand JW, Lagendijk JJW, Andersen JB, Bolomey JC (1989b) Quality assurance guidelines for E.S.H.O. protocols. In: Sugahara T, Saito M (eds) Hyperthermic oncology 1988, vol 2. Taylor and Francis, London (in press)

Harrington RF (1961) Time-harmonic electromagnetic fields. McGraw-Hill, New York

Harrington RF (1968) Field computation by moment methods. Macmillan, New York

Hartsgrove G, Kraszewski A, Surowiec A (1987) Simulated biological materials for electromagnetic radiation absorption studies. Bioelectromagnetics 8:29–36

Hazzard DG (ed) (1977) Symposium on the biological effects and measurements of radio frequency/microwaves. HEW publication (FDA) 77-8026. United States Department of Health, Education and Welfare, Rockville, MD

Hill SC, Christensen DA, Durney CH (1983) Power deposition patterns in magnetically-induced hyperthermia: a two dimensional low frequency numerical analysis. Int J Radiat Oncol Biol Phys 9:893–904

Hiraoka M, Jo S, Takahashi M, Abe M (1985) Effectiveness of RF capacitive heating in the heating of human deep-seated tumors. In: Abe M, Takahashi M, Sugahara T (eds) Hyperthermia in cancer therapy. Mag Bros, Tokyo, pp 98–99

Ho HS (1979) Design of aperture sources for deep heating using electromagnetic energy. Health Phys 37:743–750

Ho HS, Guy AW, Sigelmann RA, Lehmann JF (1971) Microwave heating of simulated human limbs by aperture sources. IEEE Trans Microwave Theory Tech MTT 19:224–231

Hopfer S (1955) The design of ridge waveguides. IRE Trans Microwave Theory Tech MTT 3:20–29

Hosmer (1928) Heating effects observed in a high-frequency static field. Science 68:325–327

Howard GCW, Sathiaseelan V, King GA, Dixon AK, Anderson A, Bleehen NM (1986) Regional hyperthermia for extensive pelvic tumours using an annular phased array applicator: a feasibility study. Br J Radiol 59:1195–1201

Hudson AC (1957) Matching the sides of a parallel plate region. IRE Trans Microwave Theory Tech MTT 5:161–162

Ikeda H, Fujii M, Sakamoto K, Kanai H (1988) RF inductive hyperthermia for deep seated tumor. In: Koga S (ed) Hyperthermic oncology in Japan '87. Imai, Yonago, Japan, pp 169–170

IRPA (1984) Interim guidelines on limits of exposure to radiofrequency electromagnetic fields in the frequency range from 100 kHz to 300 GHz (International Non-ionising Radiation Committee of the International Radiation Protection Association). Health Phys 46:975–984

Iskander MF (1982) Physical aspects and methods of hyperthermia production by rf currents and microwaves. In: Nussbaum GH (ed) Physical aspects of hyperthermia. American Institute of Physics, New York, pp 151–191

Iskander MF, Khoshdel-Milani O (1984) Numerical calculations of the temperature distribution in realistic cross sections of the human body. Int J Radiat Oncol Biol Phys 10: 1907–1912

Iskander MF, Turner PF, Dubow JB, Kao J (1982) Two dimensional technique to calculate the EM power deposition pattern in the human body. J Microwave Power 17:175–185

Jain RK, Ward-Hartley K (1984) Tumor blood flow – characterization, modifications and role in hyperthermia. IEEE Trans Sonics Ultrason SU-31:504–526

James JR, Hall PS, Wood C (1981) Microstrip antenna theory and design. Peregrinus, Stevenage, UK

James JR, Johnson RH, Henderson A (1986) Compact microwave applicators. In: Hand JW, James JR (eds) Physical techniques in clinical hyperthermia. Research Studies, Letchworth, pp 149–209

Johnk CTA (1975) Engineering electromagnetic fields and waves. Wiley, New York, chap 8

Johnson CC (1965) Field and wave electrodynamics. McGraw-Hill, New York, pp 195–202

Johnson CC, Guy AW (1972) Nonionizing electromagnetic wave effects in biological materials and systems. Proc IEEE 60:692–718

Johnson CC, Shore ML (eds) (1976) Biological effects of electromagnetic waves (vol 1, 2). HEW publications (FDA) 77-8010 and 77-8011. United States Department of Health, Education and Welfare, Rockville, MD

Johnson RH (1986) New type of compact electromagnetic applicator for hyperthermia in the treatment of cancer. Proc IEE 22:591–593

Johnson RH, James JR, Hand JW, Hopewell JW, Dunlop PRC, Dickinson RJ (1984) New low-profile applicators for local heating of tissues. IEEE Trans Biomed Eng BME 31:28–37

Johnson RH, Andrasic G, Smith DLM, James JR (1985) Field penetration of arrays of compact applicators in localized hyperthermia. Int J Hyperthermia 1:321–326

Johnson RH, Preece AW, Hand JW, James JR (1987) A new type of lightweight low-frequency electromagnetic hyperthermia applicator. IEEE Trans Microwave Theory Tech MTT 35:1317–1321

Jones CH, Carnochan P (1986) Infrared thermography and liquid crystal plate thermography. In: Hand JW, James JR (eds) Physical aspects of clinical hyperthermia. Research Studies, Letchworth, pp 507–547

Jordan EC, Balmain KG (1968) Electromagnetic waves and radiating systems, 2nd edn. Prentice-Hall, Englewood Cliffs, NJ

Jouvie F, Bolomey JC, Gaboriaud G (1986) Discussion of capabilities of microwave phase arrays for hyperthermia treatment of neck tumors. IEEE Trans Microwave Theory Tech MTT 34:495–501

Kantor G, Moon CY (1983) The performance of inductive shortwave diathermy applicators. In: IEEE MTT-S international microwave symposium digest. IEEE, New York, pp 456–458

Kantor G, Witters DM (1980) A 2450 MHz slab loaded direct contact applicator with choke. IEEE Trans Microwave Theory Tech MTT 28:1418–1422

Kapp DS, Fessenden P, Samulski TV, Bagshaw MA, Cox RS, Lee ER, Lohrbach AW, Meyer JL, Prionas SD (1988) Stanford University institutional report. Phase I evaluation of equipment for hyperthermia treatment of cancer. Int J Hyperthermia 4:75–115

Kastner R, Mittra R (1983) A new stacked two-dimensional spectral iteration technique (SIT) for analyzing microwave power deposition in biological media. IEEE Trans Microwave Theory Tech MTT 31:898–904

Kato H, Ishida T (1983) A new inductive applicator for hyperthermia. J Microwave Power 18:331–336

Kato H, Ishida T (1987) Development of an agar phantom adaptable for simulation of various tissues in the range 5–40 MHz. Phys Med Biol 32:221–226

Kato H, Hiraoka M, Nakajima T, Ishida T (1985) Deep-heating characteristics of an RF capacitive heating device. Int J Hyperthermia 1:15–28

Kato K, Matsuda J, Saito Y, Yamashita T, Hashida I, Tomaru T, Uchida I, Onai Y (1988) Computer simulation of temperature distribution in human body heated by RF capacitive hyperthermia. In: Koga S (ed) Hyperthermic oncology in Japan '87. Imai, Yonago, Japan, pp 149–150

King RWP, Smith GS (1981) Antennas in matter: fundamentals, theory and applications. MIT Press, Cambridge

Keinstein BH (1987) Biological effects of nonionizing electromagnetic radiation – a digest of current literature (produced for Office of Naval Research). Information Ventures, Philadelphia

Knoechel R (1983) Capabilities of multiapplicator systems for focused hyperthermia. IEEE Trans Microwave Theory Tech MTT 31:70–73

Knudsen M, Hartmann U (1986) Optimal temperature control with phased array hyperthermia system. IEEE Trans Microwave Theory Tech MTT 34:597–603

Koga S (ed) (1988) Hyperthermic oncology in Japan '87. Imai, Yonago, Japan

Krusen FH, Herrick JF, Leden U, Wakim KG (1947) Preliminary report of experimental studies of the heating effect of microwaves (radar) in living tissue. Proc Staff Meet Mayo Clin 22:209–224

Lagendijk JJW (1983) A new coaxial TEM radiofrequency/microwave applicator for non-invasive deep-body hyperthermia. J Microwave Power 18:367–375

Lagendijk JJW, Nilsson P (1985) Hyperthermia dough: a fat and bone equivalent phantom to test microwave/radiofrequency hyperthermia heating systems. Phys Med Biol 30:709–712

Lau RWM (1986) Computer modelling: the design and evaluation of microwave and radiofrequency hyperthermia applicators. PhD thesis, University of London

Lau RWM, Sheppard RJ (1986) The modelling of biological systems in three dimensions using the time domain finite-difference method. I. The implementation of the technique. Phys Med Biol 31:1247–1256

Lau RWM, Sheppard RJ, Howard G, Bleehen NM (1986) The modelling of biological systems in three dimensions using the time domain finite difference method. II. The application and experimental evaluation of the method in hyperthermia applicator design. Phys Med Biol 31:1257–1266

Leden UN, Herrick JF, Wakim KG, Krusen FH (1947) Preliminary studies on the heating and circulating effects of microwaves (radar). Br J Phys Med 10:177–184

Lee ER, Tarczy-Hornoch P, Fessenden P, Kapp D, Prionas S (1988) Scanning dipole antenna array applicator (Abstract Bc-6). In: Abstracts of 8th NAHG meeting, Philadelphia, April 1988. Radiation Research Society, Philadelphia, p 15

Lehmann JF, McMillan JA, Brunner GD, Johnston VC (1962) Heating patterns produced in specimens by microwaves of the frequency 2456 MHz when applied by A, B and C directors. Arch Phys Med 43:538–546

Lehmann JF, Johnston VC, McMillan JA, Silverman DR, Brunner GD, Rathbun LA (1965) Comparison of deep heating by microwaves at frequencies 2456 and 900 megacycles. Arch Phys Med 46:307–314

Lehmann JF, Guy AW, deLateur BJ, Stonebridge JB, Warren CG (1968) Heating patterns produced by shortwave diathermy using helical induction coil applicators. Arch Phys Med 49:193–198

Lehmann JF, deLateur BJ, Stonebridge JB (1969) Selective muscle heating by shortwave diathermy with a helical coil. Arch Phys Med 50:117–132

Lehmann JF, Guy AW, Stonebridge JB, deLateur B (1978) Evaluation of a therapeutic direct-contact 915 MHz microwave applicator for effective deep heating in humans. IEEE Trans Microwave Theory Tech MTT 26:556–563

Lehmann JF, Stonebridge JB, Wallace JE, Warren CG, Guy AW (1979) Microwave therapy: stray radiation, safety and effectiveness. Arch Phys Med Rehabil 60:578–584

Lehmann JF, McDougall JA, Guy AW, Chou CK, Esselman PC, Warren CG (1983) Electrical discontinuity of tissue substitute models at 27.12 MHz. Bioelectromagnetics 4:257–265

Lerch IA, Kohn S (1983) Radiofrequency hyperthermia: the design of coil transducers for local heating. Int J Radiat Oncol Biol Phys 9:939–948

Licht S (1965) History of therapeutic heat. In: Licht S (ed) Therapeutic heat and cold, 2nd edn. Licht, New Haven, pp 196–231

Lin JC (1986) Engineering and biophysical aspects of microwave and radiofrequency radiation. In: Watmough DJ, Ross WM (eds) Hyperthermia. Blackie, Glasgow, pp 42–75

Lin JC (ed) (1988) Electromagnetic interaction with biological systems. Plenum, New York

Livesay DE, Chen K (1974) Electromagnetic fields induced inside arbitrary shaped biological bodies. IEEE Trans Microwave Theory Tech MTT 22:1273–1280

Loane J, Ling H, Wang BF (1986) Experimental investigation of a retrofocusing microwave hyperthermia applicator: conjugate-field matching scheme. IEEE Trans Microwave Theory Tech MTT 34:490–494

Lumori MLD (1988) Microwave power deposition in bounded and inhomogeneous lossy media. Ph D thesis, University of Arizona

Massoudi H, Durney CH, Iskander MF (1984) Limitations of the cubical block model of man in calculating SAR distributions. IEEE Trans Microwave Theory Tech MTT 32:746–752

Marha K, Musil J, Tuha H (1971) Electromagnetic fields and the life environment. San Francisco Press, San Francisco

Matsuda T, Sugiyama A, Nakata Y (1984) Fundamental and clinical studies of radiofrequency hyperthermia and radiation therapy. In: Overgaard J (ed) Hyperthermic oncology 1984, vol 1. Taylor and Francis, London, pp 349–352

Melek M, Anderson AP (1981) A thinned cylindrical array for focused microwave hyperthermia. In: Proceedings of 11th European microwave conference. Microwave Exhibitions and Publishers, Tunbridge Wells, pp 427–432

Michaelson SM (1982) Bioeffects of high frequency currents and electromagnetic radiation. In: Lehmann JF (ed) Therapeutic heat and cold, 3rd edn. Williams and Wilkins, Baltimore, pp 278–352

Mittlemann E, Osborne SL, Coulter JS (1941) Short wave diathermy power absorption and deep tissue temperature. Arch Phys Ther 22:133–139

Morand A, Bolomey JC (1987a) A model for impedance determinations and power deposition characterization in three-electrode configurations for capacitive radio frequency hyperthermia – part A: impedance determinations. IEEE Trans Biomed Eng BME 34:217–222

Morand A, Bolomey JC (1987b) A model for impedance determinations and power deposition characterization in three-electrode configurations for capacitive radio frequency hyperthermia – part B: current flow and power deposition. IEEE Trans Biomed Eng BME 34:223–232

Morita N, Bach Andersen J (1982) Near field absorption in a circular cylinder from electric and magnetic line sources. Bioelectromagnetics 3:253–274

Mortimer B, Oshborne SL (1935) Tissue heating by short wave diathermy. JAMA 104:1413–1419

Nagelschmidt F (1913) Lehrbuch der Diathermie. Berlin

Nagelschmidt F (1928) Eine neue Methode der Wärme – Anwendung durch Diathermie. Dtsch Med Wochenschr 54: 2102–2104

NCRP (1981) Radiofrequency electromagnetic fields. Report no 67. National Council on Radiation Protection and Measurements, Bethesda

NCRP (1986) Biological effects and exposure criteria for radiofrequency electromagnetic fields. Report no 86. National Council on Radiation Protection and Measurements, Bethesda

Neymann CA (1934) Discussion. Arch Phys Ther X-Ray Radiat 15:166

Neymann CA (1938) Artificial fever. Balliere, Tindall and Cox, London

Neymann CA, Osborne SL (1929) Artificial fever produced by high frequency currents. A preliminary report. Ill Med J 56:199–203

Nikawa Y, Kikuchi M, Mori S (1985) Development and testing of a 2450 MHz lens applicator. IEEE Trans Microwave Theory Tech MTT 33:1212–1216

Nilsson P (1984) Physics and technique of microwave-induced hyperthermia in the treatment of malignant tumours. Ph D thesis, University of Lund

Nilsson P, Larsson T, Persson B (1985) Absorbed power distributions from two tilted waveguide applicators. Int J Hyperthermia 1:29–43

NRPB (1986) Advice on the protection of workers and members of the public from the possible hazards of electric and magnetic fields with frequencies below 300 GHz: a consultative document. National Radiological Protection Board, Didcot, UK

Nussbaum GH, Sidi J, Rouhanizadeh N, Morel P, Jasmin C, Convert G, Mabire JP, Azam G (1986) Manipulation of central axis heating patterns with a prototype, three-electrode capacitive device for deep-tumor hyperthermia. IEEE Trans Microwave Theory Tech MTT 34:620–625

Ohguchi Y, Tsutsumi S (1988) A CAD system of hyperthermia and its application to RF capacitive type heating of upper abdomen. In: Koga S (ed) Hyperthermic oncology in Japan '87. Imai, Yonago, Japan, pp 143–144

Oleson JR (1984) A review of magnetic induction methods for hyperthermia treatment of cancer. IEEE Trans Biomed Eng BME 31:91–97

Oleson JR, Cetas TC (1982) Clinical hyperthermia with RF currents. In: Nussbaum GH (ed) Physical aspects of hyperthermia. American Institute of Physics, New York, pp 280–306

Oleson JR, Cetas TC, Corry PM (1983a) Hyperthermia by magnetic induction: experimental and theoretical results for coaxial coil pairs. Radiat Res 95:175–186

Oleson JR, Heusenkveld RS, Manning MR (1983b) Hyperthermia by magnetic induction: clinical experience with concentric electrodes. Int J Radiat Oncol Biol Phys 9: 549–556

Oleson JR, Sim DA, Conrad J, Fletcher AM, Gross EJ (1986) Results of a phase I regional hyperthermia device evaluation: microwave annular array versus radiofrequency induction coil. Int J Hyperthermia 2:327–336

Osepchuk J (ed) (1983) Biological effects of electromagnetic radiation. IEEE, New York

Overmyer KM, Pearce JA, DeWitt DP (1979) Measurements of temperature distributions at electrosurgical dispersive electrode sites. Trans ASME J Biomech Eng 101:66–72

Paglione R, Sterzer F, Mendecki J, Friedenthal E, Botstein C (1981) 27 MHz ridged waveguide applicators for localised hyperthermia treatment of deep seated malignant tumours. Microwave J 24:71–80

Paulsen KD (1989) Calculation of power deposition patterns in hyperthermia. In: Gauthierie M (ed) Thermal modeling and thermal dosimetry. Springer, Berlin Heidelberg New York (Clinical thermology, vol 5)

Paulsen KD, Strohbehn JW, Hill SC, Lynch DR, Kennedy FE (1984a) Theoretical temperature profiles for concentric coil induction heating devices in a two-dimensional axi-symmetric, inhomogeneous patient model. Int J Radiat Oncol Biol Phys 10:1095–1107

Paulsen KD, Strohbehn JW, Lynch DR (1984b) Theoretical temperature distributions produced by an annular phased array type system in CT-based patient models. Radiat Res 100:536–552

Paulsen KD, Strohbehn JW, Lynch DR (1985) Comparative theoretical performance for two types of regional hyperthermia systems. Int J Radiat Oncol Biol Phys 11:1659–1671

Paulsen KD, Lynch DR, Strohbehn JW (1988a) Three-dimensional finite, boundary, and hybrid element solutions of the Maxwell equations for lossy dielectric media. IEEE Trans Microwave Theory Tech MTT 36:682–693

Paulsen KD, Strohbehn JW, Lynch DR (1988b) Theoretical electric field distributions produced by three types of regional hyperthermia devices in a three-dimensional homogeneous model of man. IEEE Trans Biomed Eng BME 35:36–45

Plancot M (1983) Contribution à l'étude théorique, expérimentale et clinique de l'hyperthermie microonde controlée par radiométrie microonde. Thesis, University of Science and Technology, Lille

Presman AS (1970) Electromagnetic fields and life. Plenum, New York

Rasmark P, Bach Andersen J (1984) Focused electromagnetic heating of muscle tissue. IEEE Trans Microwave Theory Tech MTT 32:887–888

Robillard M, N'Guyen DD, Chive M, Leroy Y, Audet J, Bolomey JC, Pichot C (1980) Profondeur de pénétration et résolution spatiale de sondes atraumatiques utilisées en microondes. In: Berteaud AJ, Servantie B (eds) Proceedings URSI symposium Ondes electromagnétiques et biologie, Jouy-en-Josas, July 1980. URSI, CNFRS, Thiais, pp 213–217

Roemer RB (1988) Heat transfer in hyperthermia treatments: basic principles and applications. In: Paliwal B (ed) Proceedings of 1987 American Association of Physicists in Medicine summer school on physical aspects of hyperthermia. American Institute of Physics, New York, pp 210–242

Ruggera PS (1980) Measurement of emission levels during microwave and shortwave diathermy treatments. HHS publication (FDA) 80-8119. United States Department of Health and Human Services. Bureau of Radiological Health, Rockville, MD

Ruggera PS, Kantor G (1984) Development of a family of RF helical coil applicators which produce transversely uniform axially distributed heating in cylindrical fat-muscle phantoms. IEEE Trans Biomed Eng BME 31:98–106

Ryan TP, Coughlin CT, Strohbehn JW (1988) SAR evaluation of three spiral applicator designs for 433 MHz microwave hyperthermia with clinical temperature correlation (Abstract Bc-7). In: Abstracts 8th NAHG meeting, Philadelphia, April 1988. Radiation Research Society, Philadelphia, p 15

Samulski TV, Kapp DS, Fessenden P, Lohrback A (1987) Heating deep seated eccentrially located tumors with an annular phased array system: a comparative clinical study using two annular array operating configurations. Int J Radiat Oncol Biol Phys 13:83–94

Sandhu TS, Kolozsvary A (1984) Conformal hyperthermia applicators. In: Overgaard J (ed) Hyperthermic oncology 1984, vol 1. Taylor and Francis, London, pp 675–678

Sapozink MD, Gibbs FA, Gates KS, Stewart JR (1984) Regional hyperthermia in the treatment of clinically advanced, deep seated malignancy: results of a pilot study employing and annular array applicator. Int J Radiat Oncol Biol Phys 10:775–786

Sapozink MD, Gibbs FA, Thomson JW, Stewart JR (1985) A comparison of deep regional hyperthermia from an annular phased array and a concentric coil in the same patients. Int J Radiat Oncol Biol Phys 11:179–190

Sapozink MD, Gibbs FA, Egger MJ, Stewart JR (1986) Abdominal regional hyperthermia with an annular phased array. J Clin Oncol 4:775–783

Sapozink MD, Gibbs FA, Gibbs P, Stewart JR (1988) Phase I evaluation of hyperthermia equipment – University of Utah institutional report. Int J Hyperthermia 4:117–132

Sathiaseelan V, Iskander MF, Howard GCW, Bleehen NM (1986) Theoretical analysis and clinical demonstration of the effect of power control using the annular phased-array hyperthermia system. IEEE Trans Microwave Theory Tech MTT 34:514–519

Schaubert DH (1984) Electromagnetic heating of tissue-equivalent phantoms with thin, insulating partitions. Bioelectromagnetics 5:221–232

Schereschewsky JW (1928) The action of currents of very high frequency upon tissues and cells. A: upon a transplantable mouse sarcoma. Public Health Rep 43:927–939

Schereschewsky JW (1933) Biological effects of very high frequency electro-magnetic radiation. Radiology 20:246–253

Schliephake E (1935) Short-wave diathermy. Actinic, London

Schwan HP (1957) Electrical properties of tissues and cells. Adv Biol Med Phys 5:147–209

Schwan HP (1959) Alternating current spectroscopy of biological substances. Proc IRE 47:1841–1855

Sedillot C (1853) Traite de médécine opératoire. Paris

Seguin H (1983) Progress in standardisation in the field of microwaves and radiowaves by the commission and member states of the European Community. Commission of the European Communities, document no 3501/EN. CEC, Luxembourg

Sekins KM, Emery AF (1982) Thermal science for physical medicine. In: Lehmann JF (ed) Therapeutic heat and cold, 3rd edn. Williams and Wilkins, Baltimore, pp 70–132

Shimm DS, Cetas TC, Oleson JR, Cassady JR, Sim DA (1988) Clinical evaluation of hyperthermia equipment: the University of Arizona institutional report for the NCI hyperthermia equipment evaluation contract. Int J Hyperthermia 4:39–51

Skolnik MI, King DD (1964) Self-phasing array antennas. IEEE Trans Antennas Propag AP 12:142–149

Sliney DH, Wolbarsht ML, Muc AM (1985) Differing radiofrequency standards in the microwave region – implications for future research. Health Phys 49:677–683

Soiland A (1928) Thermogenesis by radiofrequency currents. Acta Radiol 9:474–477

Song CW, Rhee JG, Lee KKL, Levitt SH (1986) Capacitive heating of phantom and human tumors with an 8 MHz radiofrequency applicator (Thermotron RF-8). Int J Radiat Oncol Biol Phys 12:365–372

Spiegel RJ (1984) A review of numerical models for predicting the energy deposition and resultant thermal response of humans exposed to electromagnetic fields. IEEE Trans Microwave Theory Tech MTT 32:730–746

Stauffer PR, Hevezi JM (1982) Possible hazards of patient anaesthesia during hyperthermia therapy. Int J Radiat Oncol Biol Phys 8:1077

Storm FK, Harrison WH, Elliott RS, Kaiser LR, Silberman AW, Morton DL (1981) Clinical radiofrequency hyperthermia by magnetic loop induction. J Microwave Power 16:179–184

Sratton JA (1941) Electromagnetic theory. McGraw-Hill, New York, chap 9

Strohbehn JW, Paulsen KD, Lynch DR (1986) Use of the finite element method in computerized thermal dosimetry. In:

Hand JW, James JR (eds) Physical techniques in clinical hyperthermia. Research Studies, Letchworth, pp 383–451

Stuchly MA, Repacholi MH, Lecuyer DW, Mann RD (1982) Exposure to the operator and patient during short wave diathermy treatments. Health Phys 42:341–366

Stuchly MA, Repacholi MH, Lecuyer DW (1983) Operator exposure to radiofrequency fields near a hyperthermia device. Health Phys 45:101–107

Stuchly SS (ed) (1979) Symposium on electromagnetic fields in biology. International Microwave Power Institute, Edmonton, Canada

Sullivan DM, Borup DT, Gandhi OP (1987) Use of the finite-difference-time-domain method in calculating EM absorption in human tissues. IEEE Trans Biomed Eng BME 34:148–157

Sullivan DM, Gandhi OP, Taflove A (1988) Use of the finite-difference time domain method for calculating EM absorption in man models. IEEE Trans Biomed Eng BME 35:179–186

Sultran MF, Mittra R (1985) An iterative moment method for analyzing the electromagnetic field distribution inside inhomogeneous lossy dielectric objects. IEEE Trans Microwave Theory MTT 33:163–168

Taflove A, Brodwin ME (1975a) Numerical solution of steady state electromagnetic problems using the time-dependent Maxwell's equations. IEEE Trans Microwave Theory Tech MTT 23:623–630

Taflove A, Brodwin ME (1975b) Computation of the electromagnetic fields and induced temperatures within a model of the microwave-irradiated human eye. IEEE Trans Microwave Theory Tech MTT 23:888–896

Takahashi Y, Nikawa Y, Mori S, Nakagawa M, Kikuchi M (1985) Electromagnetic field convergent applicator for microwave hyperthermia at 433 MHz. In: Abe M, Takahashi M, Sugahara T (eds) Hyperthermia and cancer therapy. Mag Bros, Tokyo, pp 132–133

Tanabe E, McEuen AH, Norris CS, Fessenden P, Samulski TV (1983) A multielement microstrip antenna for local hyperthermia. In: IEEE MTT-S International microwave symposium digest (IEEE 83 CH 1871-3). IEEE, New York, pp 183–185

Tofani S, Agnesod G (1984) The assessment of unwanted radiation around diathermy RF capacitive applicators. Health Phys 47:235–241

Tsai CT, Massoudi H, Durney CH, Iskander MF (1986) A procedure for calculating fields inside arbitrarily shaped, inhomogeneous dielectric bodies using linear basis functions with the moment method. IEEE Trans Microwave Theory Tech MTT 34:1131–1139

Turner PF (1983) Electromagnetic hyperthermia devices and methods. MS Thesis, University of Utah, Salt Lake City, Chap 2

Turner PF (1984a) Hyperthermia and inhomogeneous tissue effects using an annular phased array. IEEE Trans Microwave Theory Tech MTT 32:874–882

Turner PF (1984b) Regional hyperthermia with an annular phased array. IEEE Trans Biomed Eng BME 31:106–114

Turner PF (1986) Mini-annular phased array for limb hyperthermia. IEEE Trans Microwave Theory Tech MTT 34:508–513

Turner PF (1988) Operational and clinical aspects of the BSD-2000. Proceedings of Essen University and BSD Medical Corporation symposium, Essen, April 1988

Turner PF, Kumar L (1982) Computer solution for applicator heating pattern. NCI Monogr 61:521–523

USSR (1976) Occupational safety standards system. Electromagnetic fields of radiofrequency. General safety requirements. GOST Standard 12.1.006-76. Gosudarstvennyii Komitet Standardov Sovieta Ministriv USSR, Moscow

USSR (1984) Occupational safety standards system. Electromagnetic fields of radiofrequency, permissible levels in work places and requirements for control. GOST Standard 12.1.006-84. Gosudarstvennyii Komitet Standardov Sovieta Ministriv USSR, Moscow

van den Berg PM (1984) Iterative computational techniques in scattering based upon the integrated square error criterion. IEEE Trans Antennas Propag AP 32:1063–1071

van den Berg PM, de Hoop AT, Segal A, Praagman N (1983) The computational model of the electromagnetic heating of biological tissue with application to hyperthermic cancer therapy. IEEE Trans Biomed Eng BME 30:797–805

van Koughnett AL, Wyslouzil W (1972) A waveguide TEM mode exposure chamber. J Microwave Power 7:381–383

van Putten MHPM, van den Berg PM (1986) A three-dimensional model for the 'coaxial TEM' deep body hyperthermia applicator. Int J Hyperthermia 2:243–252

van Rhoon GC, Visser AG, van den Berg PM, Reinhold HS (1984) Temperature depth profiles obtained in muscle equivalent phantoms using the RCA 27 MHz ridged waveguide. In: Overgaard J (ed) Hyperthermic oncology 1984 vol 1. Taylor and Francis, London, pp 499–502

van Rhoon GC, Visser AG, van den Berg PM, Reinhold HS (1988) Evaluation of ring capacitor plates for regional deep heating. Int J Hyperthermia 4:133–142

Vaughan R, Bach Andersen J (1985) Polarization properties of the helix antenna. IEEE Trans Antennas Propag AP 33:10–20

Visser AG, van Rhoon GC, van den Berg PM, Reinhold HS (1987) Evaluation of calculated temperature distributions for a 27 MHz ridged waveguide used in localized deep hyperthermia. Int J Hyperthermia 3:245–256

von Ardenne M, von Ardenne T, Bohme G, Reitnauer PG (1977) Selektive Lokalhyperthermie der Krebsgewebe. Homogenisierte Energiezufuhr auch in tief liegende Gewebe der Hochleistungs-Dekawellen-Spulenfeld + Rasterbewegung des Doppelsystems. Arch Geschwulstforsch 47:487–523

von Zeynek RR, von Bernd E, von Preyss W (1908) Über Thermopenetration. Wien Klin Wochenschr 21:517–520

Wait JR (1959) Electromagnetic radiation from cylindrical structures. Pergamon, London

Wait JR (1985) Focused heating in cylindrical targets: part I. IEEE Trans Microwave Theory Tech MTT 33:647–649

Wait JR (1986) Analysis of the radiation leakage for a four-aperture phased array applicator in hyperthermia therapy. IEEE Trans Microwave Theory Tech MTT 34:531–541

Wait JR, Lumori MLD (1986) Focussed heating in cylindrical targets: part II. IEEE Trans Microwave Theory Tech MTT 34:357–359

Wang CQ, Gandhi OP (1989) Numerical simulation of annular phased arrays for anatomically based models using the FDTD method. IEEE Trans Microwave Theory Tech. MTT 37:118–126

Wiley JD, Webster JG (1982) Analysis and control of the current distribution under circular dispersive electrodes. IEEE Trans Biomed Eng BME 29:381–385

Wilsey TR, McEuen AH, Fessenen P, Lee ER, Tanabe E, Nelson LV, Schlitter RC, Kapp DS (1988) Arm cuff microwave microstrip array applicator (Abstract Bc-5). In: Abstracts for 8th NAHG meeting, Philadelphia, April 1988. Radiation Research Society, Philadelphia, p 15

Wong TZ, Strohbehn JW, Douple EB (1985) Automated measurement of power deposition patterns from interstitial microwave antennas used in hyperthermia. In: Kuklinski WS, Ohley WJ (eds) Proceedings 11th Northeast Bioengineering Conference. IEEE, New York, pp 58–61

Yamashita E, Mittra R (1968) Variational method for the analysis of microstrip lines. IEEE Trans Microwave Theory Tech MTT 16:251–256

Yee KS (1966) Numerical solution of initial boundary value problems involving Maxwell's equations in isotropic media. IEEE Trans Antennas Propag AP 17:585–589

2 Biophysics and Technology of Ultrasound Hyperthermia

K. HYNYNEN

2.1 Introduction

Although characterization of tissues with ultrasound has been a subject of wide interest for over 20 years, there has not been a similar interest in using ultrasound in cancer therapy since the early trials in the 1930s, when ultrasound was used in a manner similar to the use of X-rays for therapeutic purposes. There are probably several reasons for this. First, the theory and equipment used in the field of diagnostic ultrasonics also have other applications, e.g., in defense (sonar) and in industry (flaw detection). Therefore, there are more resources and personnel available for research and development. Second, the therapeutic effects of ultrasound cannot be quantified by measuring the intensity of the beam (as is the case with X-rays), but by the temperature elevation induced in the tumor, which produces the beneficial effects. Interest was further reduced by the rapid development of radiotherapy as a method of treating tumors. It was not until the end of the 1970s that the potential of ultrasound as a method of inducing hyperthermia was shown (Fig. 2.1). Even since then it has not become as popular as microwaves, despite its many advantages. The main reason for this lack of popularity has been that the devices required to utilize ultrasound properly in tumor heating are fairly complex and have not yet become commercially available. If good devices do become available, it is expected that there will be increased interest in ultrasound hyperthermia.

The early studies mentioned above have been reviewed by Kremkau (1979) in detail; thus, they are only briefly summarized here. In the first study, ultrasound exposure did not have any effect on tumor growth (Szent-Gyorgyi 1933). The next report (Nakahara and Kobayashi 1934) indicated that the growth of mouse adenocarcinoma was actually stimulated by small periods of sonication (frequency 0.5 MHz, intensity $2 \ W \ cm^{-2}$, and length of sonication 60 s). Later, subsequent papers indicated that longer treatment times (up to several minutes) were required to obtain beneficial effects in the tumor. This was followed by number of studies, including the first clinical test reported by Horvath (1944). Great enthusiasm followed, which allowed for a number of clinical studies in which ultrasound was used alone, with variable levels of exposure, to treat tumors. Discouraged by the poor clinical benefit obtained in these treatments, the Ultrasound in Medicine Conference held in Erlangen, Germany in 1949, stated that ultrasound was not suitable for tumor therapy and that its clinical use should be discontinued. This statement, known as the

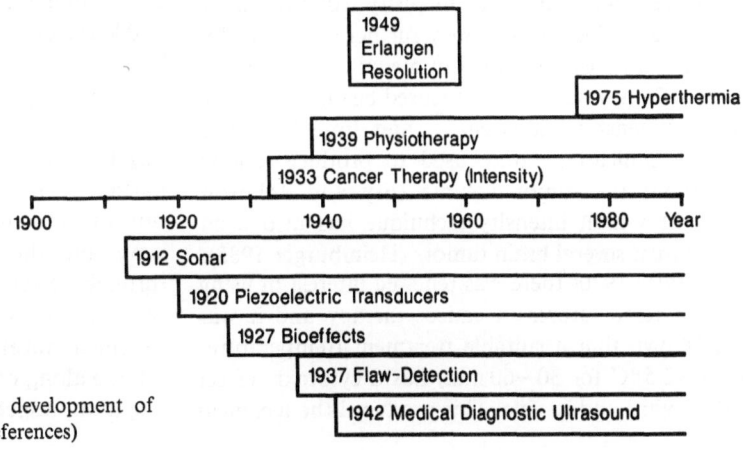

Fig. 2.1. The main historical events in the development of medical ultrasonics (see Kremkau 1979 for references)

"Erlangen Resolution," had a strong impact on the clinical use of ultrasound as a cancer therapy.

During the 1950s and 1960s a few studies investigated the combined effects of ultrasound and radiation, and encouraging results were obtained. The therapeutic effect of sonication was found to be caused by the subsequent temperature elevation in the tissue. These studies led to the conclusion that ultrasound alone should not be used but that in combination with radiotherapy its use can lead to beneficial effects (Woeber 1965; Clarke et al. 1970).

During this same period there was another interesting approach. First, as related to cancer therapy, Burov (1956) proposed a new protocol for ultrasound treatments of cancer: high intensity sonication for a period of a few seconds. The hypothesis behind this was that the high intensity beam would produce beneficial nonthermal effects. The treatment (frequency 1.5 MHz, intensity 150 W cm^{-2}, and sonication period 1–3 s) produced resorption of rabbit carcinomas and also of human malignant melanomas (Burov 1956; Burov and Andreevskaya 1956). This high intensity approach was also successfully tested in a clinical study in Japan, where it was found to destroy selectively breast tumors and several brain tumors (Oka 1960). Parallel with these studies, high intensity ultrasound was used to produce lesions in both animal and human central nervous systems for surgical purposes (Fry 1965). From these experiments it was obvious that the tissue damage was caused by the temperature elevation in the focal zone producing coagulation necrosis. The intensity threshold for the production of the lesions was investigated in several types of tissue and was found to be dependent on time in a manner similar to that of lesions produced with other heating methods. This high intensity technique, with intensity levels large enough to produce coagulation necrosis, has also been used in cancer therapy. Kishi et al. (1975) observed favorable responses in murine gliomas implanted in abdominal walls (sonication frequency 900 kHz, intensity 1000 W cm^{-2}, and time 2 s). These results were further supported by the good responses obtained in medulloblastomas in hamster flank when using a focused beam (frequency 1 MHz, intensity 720 W cm^{-2}, and time 7 s), and multiple sonications were used to produce lesions throughout the tumor volume (Fry and Johnson 1978). This high intensity technique has also been used to treat several brain tumors (Heimburger 1985).

In the early 1970s there was renewed interest in using elevated temperatures in cancer therapy and it was soon shown that a suitable treatment (temperatures above 42.5 °C for 30–60 min) has a cytotoxic effect both in vitro and in vivo. This produced the technical challenge of heating the tumors in a clinical setting: a problem that has not yet been completely solved. Since ultrasound had been used to heat nonmalignant tissue for physiotherapy purposes (Lehmann 1965), it was one of the methods investigated for the induction of hyperthermia. Lele (1975) demonstrated the feasibility of scanning focused ultrasound beam to obtain controllable hyperthermic temperatures in vivo. Marmor et al. (1977) reported positive tumor responses and cure of sarcomas and carcinomas in mice exposed to ultrasonically induced hyperthermia. Similar benefits were also demonstrated in spontaneous animal tumors (Marmor et al. 1978) and superficial human tumors (Marmor et al. 1979); in the latter there was an objective response rate of about 55% when the temperature was kept between 43° and 45 °C for 30 min. When the hyperthermia treatments were combined with radiotherapy, the response rate was increased to 80% (Corry et al. 1982). During this same period, Dr. Lele was testing the feasibility of using scanned focused ultrasound in clinical treatments. He produced therapeutic temperatures in tumors of various sizes and locations and showed that even tumors up to 12 cm deep can be heated selectively with scanned focused ultrasound (Lele 1983, 1984, 1986).

At present, ultrasound appears to offer good control over the resultant temperature distributions in the sites where it can be used. Also, it seems to be the only noninvasive method capable of inducing therapeutic temperatures in deep tumors. The main difficulty in clinically testing ultrasound hyperthermia is the lack of commercial equipment with enough flexibility to heat various tumors routinely.

2.2 Basic Physics of Ultrasound

Ultrasound is a form of vibrational energy (more than 18000 cycles/s) that is propagated as mechanical wave by the motion of particles within the medium. The wave propagation causes compressions and rarefactions of the particles, and thus a pressure wave is propagating along with the mechanical movement of the particles. The characteristics of the wave are a function of both the original disturbance generating the wave and the acoustic properties of the medium through which it travels. The propagating wave can be of two types, either longitudinal or transverse (= shear wave), depending on whether the particles vibrate along or across the direction of the wave propagation, respectively. There are also two types of surface wave. However, during an ultrasound hyperthermia treatment the propagating waves are mainly longitudinal, with the generation of shear waves only under special circumstances such as soft tissue-bone interfaces. Therefore, the theory of longitudinal waves will be reviewed here. (For further details of the theory see, for example, Hueter and Bolt 1955 and Wells 1969, 1977.)

2.2.1 Physical Aspects of Ultrasound Waves

2.2.1.1 Wave Equation

First consider a particle in a simple harmonic motion, oscillating around its rest position (Fig. 2.2). For such a motion, the *particle displacement* (ξ) relative to the origin of the coordinate system is given as a function of time by the relation:

$$\xi = \xi_a \sin(\omega t + \phi) \tag{2.1}$$

where ω is the angular frequency, the quantity of $\omega t + \phi$ is called the phase, and thus ϕ is the initial phase at $t = 0$. Now it is easy to see that the displacement varies between $+\xi_a$ and $-\xi_a$. The maximum displacement, ξ_a, is called the amplitude of the simple harmonic motion. The particle displacement repeats itself with certain time intervals as characterized by the sine function. This time interval is called the *period of motion* (τ). The frequency of the oscillations (f) is equal to the number of complete cycles per unit time, and it is:

$$f = 1/\tau \tag{2.2}$$

The *angular frequency* of the harmonic motion (ω) is:

$$\omega = 2\pi f \tag{2.3}$$

From Eq. (2.1) the *velocity of the particle* (u) is defined as:

$$u = d\xi/dt = \omega \xi_a \cos(\omega t + \phi) \tag{2.4}$$

Here we can see that the particle velocity is $\pi/2$ out of phase from the displacement. Similarly, a second differentiation yields the *particle acceleration* (a):

$$a = du/dt = -\omega^2 \xi_a \sin(\omega t + \phi) = -\omega^2 \xi \tag{2.5}$$

This shows that the acceleration is always proportional and opposite to the displacement of the particle.

Now let us consider the situation where a longitudinal wave is propagating in a medium (Fig. 2.3). Each particle will oscillate around its rest position with the frequency of the propagating wave. The *wavelength* (λ) is defined as the distance between particles that are in the same phase of motion; for example, the minimum distance between two particles at the maximum positive displacement. The wavelength can be calculated from the *propagation velocity of the wave* (v) in the medium and the *period of the wave* (τ) or the *frequency* (f):

$$\lambda = v\tau = v/f \tag{2.6}$$

Another parameter often used to describe the wave is the *wave number* (k):

$$k = 2\pi/\lambda \tag{2.7}$$

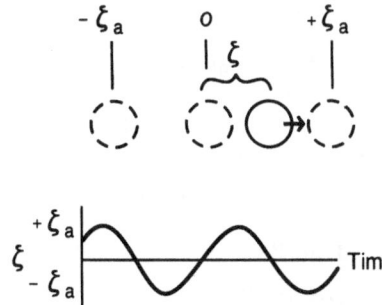

Fig. 2.2. A particle oscillating about its rest position (*top*) and its displacement associated with the motion as a function of time (*bottom*)

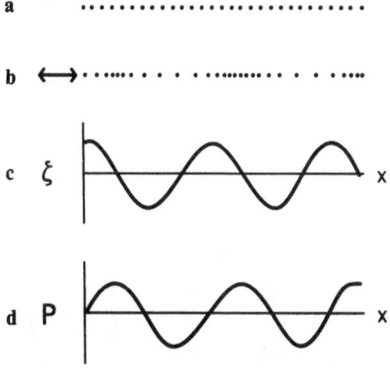

Fig. 2.3. a A row of undisturbed particles; b a longitudinal wave propagating in the particle row; c the particle displacement and d the pressure wave propagating with the mechanical wave as a function of distance

For the situation presented in Fig. 2.4, where a compressing force F is applied over the cross-sectional area A of a rod, the relationship between the *compressing force* F_x at the distance of x and the longitudinal strain $d\xi/dx$ is given by Hooke's law:

$$F_x = -AK d\xi/dx \tag{2.8}$$

where A is the cross-sectional area and K is the bulk modulus of the medium. At the distance of dx from x ($x' = x+dx$) the force (F_x') can be written as:

$$F_x' = F_x + (\partial F_x/\partial x) dx \tag{2.9}$$

and

$$dF_x = F_x' - F_x = (\partial F_x/\partial x) dx \tag{2.10}$$

Now, by differentiating Eq. 2.8 with respect to x, multiplying through by dx, it follows that:

$$(\partial F_x/\partial x) dx = -AK(\partial^2 \xi/\partial x^2) dx \tag{2.11}$$

This change in the force has been absorbed by the particles in the mass element $\varrho_0 A dx$ (ϱ_0 is the density without a disturbance). For this system, the summation of the forces acting on the particle must equal zero; thus, it follows from Newton's second law of mechanics that:

$$-AK(\partial^2 \xi/\partial x^2) dx + \varrho_0 A dx(\partial^2 \xi/\partial t^2) = 0 \tag{2.12}$$

which, upon rearrangement, becomes:

$$\partial^2 \xi/\partial t^2 = (K/\varrho_0)\partial^2 \xi/\partial x^2 \tag{2.13}$$

or

$$\partial^2 \xi/\partial t^2 = v^2 \partial^2 \xi/\partial x^2 \tag{2.14}$$

where v is the propagation velocity of the wave and is defined as:

$$v = (K/\varrho_0)^{1/2} \tag{2.15}$$

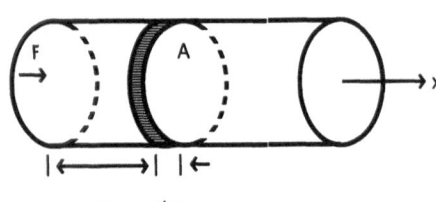

x   dx

Fig. 2.4. A longitudinal wave propagating in a rod

Equation 2.14 is called the *differential equation of a wave motion* and its solution for a simple harmonic wave propagating in the x-direction in a nonattenuating medium, is

$$\xi(x,t) = \xi_a \sin \omega(t-x/v) \tag{2.16}$$

where $\xi(x,t)$ is the particle displacement at the distance x and time t. Similarly, the particle velocity (u) can be defined as follows:

$$u(x,t) = d\xi/dt = \xi_a \omega \cos \omega(t-x/v) \tag{2.17}$$

For propagating waves in a compressible medium, the pressure changes cause the density of the medium to fluctuate in a similar manner. If p_0 and ϱ_0 are the equilibrium pressure and density, respectively, and p' and ϱ' are the same parameters for a disturbed medium, the conservation of matter for the situation in Fig. 2.4 can be written as follows:

$$\varrho' A(dx+d\xi) = \varrho_0 A dx \tag{2.18}$$

or

$$\varrho' = \varrho_0/(1+\partial\xi/\partial x) \tag{2.19}$$

Since $\partial\xi/\partial x$ is small, we can use the binomial expansion to write

$$\varrho' = \varrho_0(1-\partial\xi/\partial x) \tag{2.20}$$

The solution for the pressure p' is a function of the density change; thus, using Taylor's expansion series, it can be defined as follows:

$$p' = p_0 + (\varrho'-\varrho_0)(dp/d\varrho')_0 \\ + \tfrac{1}{2}(\varrho'-\varrho_0)_0^2(d^2p/d\varrho'^2)_0 + \ldots \tag{2.21}$$

When the density changes are small the higher order terms can be neglected and a good first order approximation can be obtained by using the first two terms in Eq. 2.21

$$p' = p_0 + K(\varrho'-\varrho_0)/\varrho_0 \tag{2.22}$$

where K is the bulk modulus of elasticity defined as $\varrho_0(dp/d\varrho)_0$. Now, using Eq. 2.20, we define the acoustic pressure p ($= p'-p_0$) as:

$$p = -K\partial\xi/\partial x \tag{2.23}$$

$$= -\varrho_0 v^2 \partial\xi/\partial x \tag{2.24}$$

By substituting the derivative of Eq. 2.16 with respect to x, we get

$$p = \varrho_0 v \xi_a \omega \cos(\omega(t - x/v)) \qquad (2.25)$$

$$= \varrho_0 v u \qquad (2.26)$$

2.2.1.2 Acoustic Impedance

The acoustic impedance (Z) of the medium is defined as the ratio of the sound pressure p to the particle velocity u at any part of the field. Thus, from Eqs. 2.26 and 2.15 it follows that:

$$Z = \varrho_0 v = (\varrho_0 K)^{1/2} \qquad (2.27)$$

An important consideration during sonication of the tissues is the reflection of ultrasound beams from the interfaces of two media with different acoustic impedance. When an ultrasound beam meets the interface of two media (from 1 to 2) it may be partly reflected and partly transmitted to the other medium (Fig. 2.5). The incident angle (θ_i) and the angle of reflection (θ_r) are equal, i.e., $\theta_i = \theta_r$. The transmission angle (θ_t) can be determined from Snell's law as follows:

$$\sin \theta_t / \sin \theta_i = v_2 / v_1 \qquad (2.28)$$

The ratio between the reflected (p_r) and the incident (p_i) acoustic pressure of the wave depends on the incident angle and the acoustic impedance of each medium. For plane waves it is defined as:

$$\frac{p_r}{p_i} = \frac{Z_2 \cos \theta_i - Z_1 \cos \theta_t}{Z_2 \cos \theta_i + Z_1 \cos \theta_t} \qquad (2.29)$$

This relation applies when the wavelength of the plane wave is smaller than the dimensions of the reflecting object. When the wavelength is comparable to or longer than the dimensions of the object, the wave is scattered in all directions. The total scattered acoustic power depends on the size, shape, and acoustic properties of the object.

2.2.1.3 Intensity

The total mechanical energy of a particle in a simple harmonic oscillation is the sum of the kinetic and potential energies. When $\xi = 0$, the particle is passing its rest position and it has only kinetic energy. Thus, the total energy of the oscillating particle (mass = M) is

$$e = M u_0^2 / 2 \qquad (2.30)$$

where is u_0 is the particle velocity amplitude. The total energy in a unit volume is the sum of the energy of all of the particles, and the total mass of the particles/unit volume is determined by the density (ϱ_0). Thus, the energy density (E = total energy/volume) can be written as:

$$E = \varrho_0 u_0^2 / 2 \qquad (2.31)$$

The rate of energy flow through a unit area normal to the direction of the wave propagation is called the *acoustic intensity* (I). It is defined as the product of the propagation velocity of the wave and the energy density:

$$I = v E = \varrho_0 v u_0^2 / 2 = p_a^2 / 2Z \qquad (2.32)$$

where p_a is the *acoustic pressure amplitude*.

2.2.1.4 Wave Propagation in a Lossy Medium

In a real medium, the ultrasound energy is attenuated according to an exponential law. For the plane wave case, assuming the intensities are not large enough to cause wave distortion, the intensity I(x) at the depth x is described by the following formula:

$$I(x) = I(0) e^{-2\mu x} \qquad (2.33)$$

where I(0) is the intensity at the surface and μ is the attenuation coefficient per unit path length (unit Npm^{-1} or m^{-1}) (Fig. 2.6). Part of the attenuated energy is absorbed (α = absorption coefficient; unit Npm^{-1}), raising the temperature of the medium, and part is scattered away from the field. This scattered

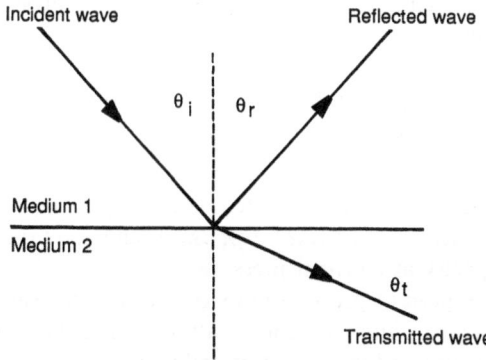

Fig. 2.5. The reflection and refraction of an ultrasonic wave at a plane surface

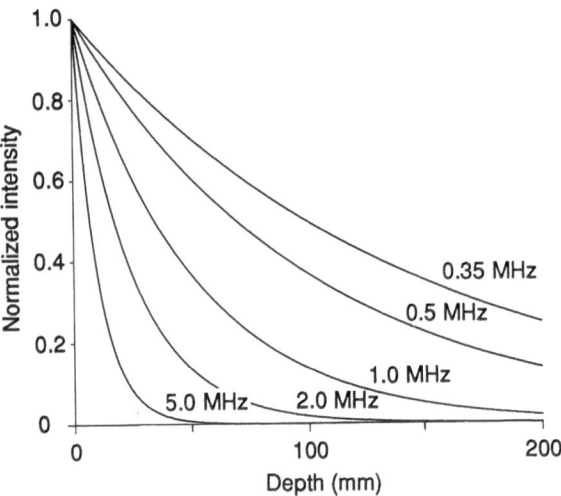

Fig. 2.6. The attenuation of an ideal plane wave while it propagates in a tissue. These calculations are based on Eq. 2.33 and an attenuation coefficient of $10 \, \text{Npm}^{-1} \, \text{MHz}^{-1}$

energy is absorbed during its propagation in tissue and thus it also contributes to the temperature elevation. From the point of view of hyperthermia treatments, almost all of the attenuated energy is absorbed in the heating field and thus the attenuation and absorption coefficients are almost equal.

Generally, for any continuous, single frequency, ultrasound field, when the effects of shear viscosity are small, the temporal average absorbed power density $\langle q \rangle$ depends on the square of the acoustic pressure amplitude p_a as follows:

$$\langle q \rangle = \alpha \, p_a^2 / \varrho_0 v \qquad (2.34)$$

where ϱ_0 is the density of the medium without the sound field.

In a plane wave situation, this can be expressed as:

$$\langle q \rangle = 2 \alpha I \qquad (2.35)$$

In the nearfield, especially with focused fields, the plane wave conditions are not satisfied (except at the focus). Thus, when these fields are used for the induction of hyperthermia, the acoustic pressure amplitude should be used to quantify the effect of the beam.

The temperature elevation induced during sonication is directly proportional to the absorbed energy $\langle q \rangle$ and depends also on the blood perfusion rate and thermal conduction of the tissue. During the initial phase of a sound pulse, the effects of blood perfusion and thermal conduction can be ignored and the rate of the temperature elevation can be written as:

$$dT/dt = \langle q \rangle / \varrho c \qquad (2.36)$$

where c is the *specific heat* of the medium. For a plane wave situation this is given as:

$$dT/dt = 2 \alpha I / \varrho c \qquad (2.37)$$

When considering focused ultrasonic fields, the thermal conduction effects become significant after a few hundred milliseconds depending on the size of the focus. In general, the larger the focal region, the smaller are the conduction effects.

2.2.1.5 Nonlinear Propagation

The previous description of the propagation of the ultrasound field required that the particle displacement amplitude was small so that the approximation of Eq. 2.21, using only the first two terms, is valid. With larger displacement amplitudes, the second order nonlinear term must be taken into account. Thus, after modifications, the *acoustic pressure* can be defined as:

$$p = (\varrho' - \varrho_0) \, (\partial p / \partial \varrho')_{0,s}$$
$$+ \tfrac{1}{2} (\varrho' - \varrho_0)^2 \, (\partial^2 p / \partial \varrho'^2)_{0,s} + \ldots \qquad (2.38)$$

or

$$p = A \, [(\varrho' - \varrho_0)/\varrho_0] + B/2 \, [(\varrho' - \varrho_0)/\varrho_0]^2 + \ldots \qquad (2.39)$$

where $A = \varrho_0 \, (\partial p / \partial \varrho')_{0,s} = \varrho_0 v_0^2$
and $B = \varrho_0^2 \, (\partial^2 p / \partial \varrho'^2)_{0,s}$

The subscripts 0 and s denote the partial derivatives at equilibrium density $(\varrho' = \varrho_0)$ and at constant entropy, s. The ratio B/A is called the second order parameter of nonlinearity and it is expressed as follows:

$$B/A = 2 \varrho_0 v_0 \, (\partial v / \partial p)_{0,s} = 2 \varrho_0 v_0 \, (\partial v / \partial p)_T$$
$$+ (2 v_0 T \gamma)/c_p \, (\partial v / \partial T)_p \qquad (2.40)$$

where T is the absolute temperature, γ is the *volume coefficient of thermal expansion*, and c_p is the heat capacity at constant pressure.

The speed of propagation v of the acoustic wave at a distance x and moment t is a function of the acoustic pressure and is defined as follows:

$$v(x, t) = v_0 + [1 + B/(2A)] u(x, t) \qquad (2.41)$$

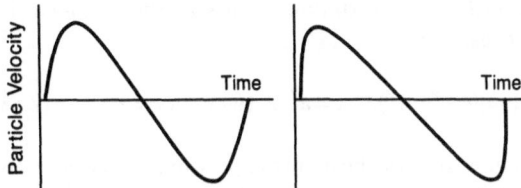

Fig. 2.7. The waveform distortion of a sinusoidal wave due to nonlinear propagation. *Left:* A significant distortion which has a small effect on the energy absorption. *Right:* Close to the shock wave, the distortion significantly increases the energy absorption

where u(x, t) is the particle velocity at that point and at that moment. It is now apparent that the speed of propagation varies, depending on the phase of the particle oscillation (and on the acoustic pressure), and is highest during the compressional phase of the wave and lowest during the rarefaction phase. This results in a cumulative distortion of the propagating wave front (Fig. 2.7), in the time domain, due to the generation of harmonic frequencies in the frequency spectrum of the wave. The amount of distortion increases as the distance from the source increases (the attenuation of the wave is ignored) until vertical discontinuities in the wave form are generated. The distortion will not progress any further in a real medium, since a particle cannot have multiple displacements at one time. This limiting wave is known as *shock*. At this point, an increase in the beam energy cannot be transmitted to a greater distance, and all of the added energy will be absorbed from the wave. Thus, the shock wave also limits the amount of energy that a mechanical wave can transmit. The local forces on the particle become very large in the shock wave.

The dependence of the wave form distortion on the acoustic pressure amplitude p_a (or intensity I), distance from the transducer (x), frequency (f), and the acoustic properties of the medium can be described by the *shock parameter*, σ, which for plane waves is defined as

$$\sigma = \beta \varepsilon\, kx \tag{2.42}$$

and for spherically focused waves as

$$\sigma = \beta \varepsilon\, k\, R \ln(R/r) \tag{2.43}$$

where R is the focal length, r is the distance from the center of curvature, $\beta = 1 + B/2A$, $k = 2\pi/\lambda$ is the wave number, and $\varepsilon = u_0/v_0$ is the acoustic Mach number, where u_0 is the particle velocity amplitude at the source. From the previous theory (Eqs. 2.26 and 2.32) one can derive the following expression:

$$\varepsilon = u_0/v_0 = p_a/(\varrho_0 v_0^2) = (2\,I/(\varrho_0 v_0^3))^{1/2} \tag{2.44}$$

The theory used in the above derivation assumes uniform pressure across a phase front. The value $\sigma = 1$ indicates the beginning of shock formation and the value $\sigma = 3$ will be obtained when the wave has reached a sawtooth form.

In a lossy medium like tissues with frequency-dependent ultrasonic attenuation and absorption, the harmonic generation causes increased attenuation since the higher frequency components will be attenuated faster than the fundamental ones. Thus, the overall attenuation and the absorption of the wave is a sum of the attenuation and absorption of each of its frequency components. For further discussion see the review of Carstensen and Muir (1986).

2.2.2 The Ultrasonic Fields

An ideal point source of ultrasonic energy emits radiation equally in all directions, resulting in a spherical wave front. The ultrasonic field from a real source can be analyzed by taking it as equivalent to a large number of point sources situated very close together, and then analyzing the resultant wave front (Huyghen's principle). The finite size of transducers causes various boundary phenomena (reflection, refraction, and diffraction), which, when combined with the inhomogeneities of the transducer material, cause a deviation of the real acoustic field from the theoretical one. However, in many cases the ultrasonic field calculated using the ideal, uniformly vibrating, transducer is a fairly good approximation of the real field generated by a transducer of the same physical dimensions and operating frequency. This is especially true for the case of the continuous wave fields used in the ultrasound hyperthermia treatments.

2.2.2.1 Unfocused Ultrasonic Fields

The acoustic pressure amplitude distribution emitted by a planar, circular transducer, oscillating as a piston (radius = a) in simple harmonic motion, is dependent on the ratio between the diameter and the wavelength (Fig. 2.8). The intensity I(x) along the central axis of propagation can be calculated from the following expression:

$$I(x) = I(0) \sin^2 [(\pi/\lambda)(\sqrt{a^2 + x^2} - x)] \tag{2.45}$$

where x represents the distance from the transducer

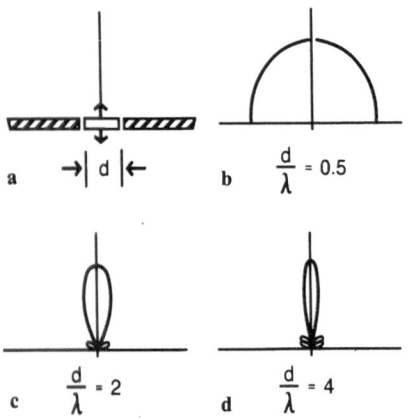

Fig. 2.8a–d. The ultrasonic field pattern emitted by a circular piston. **a** Diagram of the sonication system. **b–d** The field pattern from the different ratios of d/λ, where d is the diameter of a circular source and λ is the wavelength. (Washington 1961)

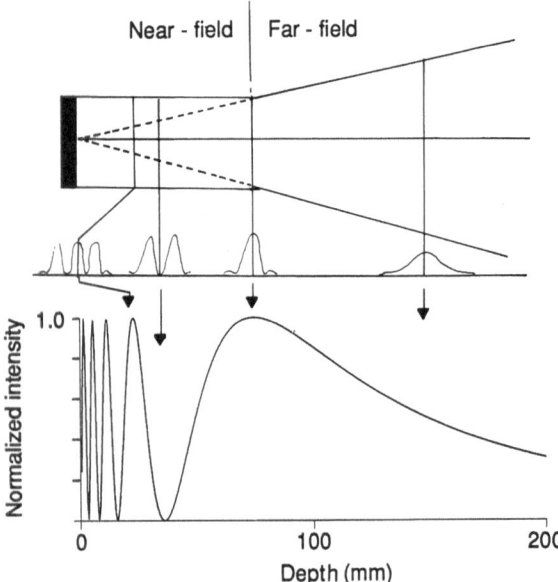

Fig. 2.9. The ultrasonic field distribution from a circular, flat transducer. *Top:* the beam outline; *middle:* the cross-sectional (pressure amplitude)^2 distributions; *bottom:* the axial (pressure amplitude)^2 distribution

and I(0) is the value of I(x) for x = 0. The intensity reaches its local maximum values at the points where:

$$x = [4a^2 - \lambda^2(2m+1)^2]/[4\lambda(2m+1)],$$
$$\text{where } (m = 0, 1, 2\ldots) \qquad (2.46)$$

and is reduced to zero when:

$$x = (a^2 - \lambda^2 m^2)/2m\lambda, \text{ where } (m = 1, 2, 3\ldots n) \qquad (2.47)$$

The last axial maximum occurs at the distance defined as:

$$x_{max} = (4a^2 - \lambda^2)/4\lambda \simeq a^2/\lambda \text{ (when } a \gg \lambda) \qquad (2.48)$$

The intensity (or the acoustic pressure amplitude) distribution cannot be analytically solved in other parts of the field. However, in general the region between the transducer and the last axial maximum (the near-field or Fresnel zone) has pressure maxima and minima rings symmetrically around the central axis, and thus the distribution of acoustic energy is not uniform. The number of maxima and minima across the beam depends upon the values of x and a/λ. Generally, the frequency of the peaks increases with decreasing x and increasing values of a/λ. As one can see from Fig. 2.9, the number of pressure maxima across the beam increases from one at the last axial maximum, to two at the last axial minimum and three at the second to last axial maximum, etc. The beam also narrows toward the last axial maximum, being about onequarter of the diameter of the transducer (−3 dB beam diameter) at the last axial maximum. The ultrasonic field beyond the last axial maximum (the far-field or Fraunhofer zone) is diverging and the intensity follows the inverse square law, I(x) α 1/x^2. The intensity distribution across the beam in the far-field can be approximated by multiplying the axial intensity by the *directivity function*, D:

$$D = 2 J_1(ka \sin\theta)/ka \sin\theta \qquad (2.49)$$

where J_1 is the Bessel function of the first kind, θ is the divergence angle, and k is the wave number ($= 2\pi/\lambda$).

Most of the plane wave ultrasound hyperthermia applicators are between 3 and 10 cm in diameter and operate between 0.5 and 5 MHz; thus, the heated region is in the near-field. At these frequencies, especially above 1 MHz, the pressure peaks and valleys are so close to each other that the thermal conduction smooth the temperature distribution. If the ultrasonic crystal is properly mounted and functioning correctly, the average energy distribution should be fairly uniform and cover almost the whole surface area of the applicator.

2.2.2.2 Focused Ultrasonic Fields

The shape of an ultrasound beam emitted by a transducer can be modified by focusing. In a fashion similar to those used in optics, the ultrasonic beams

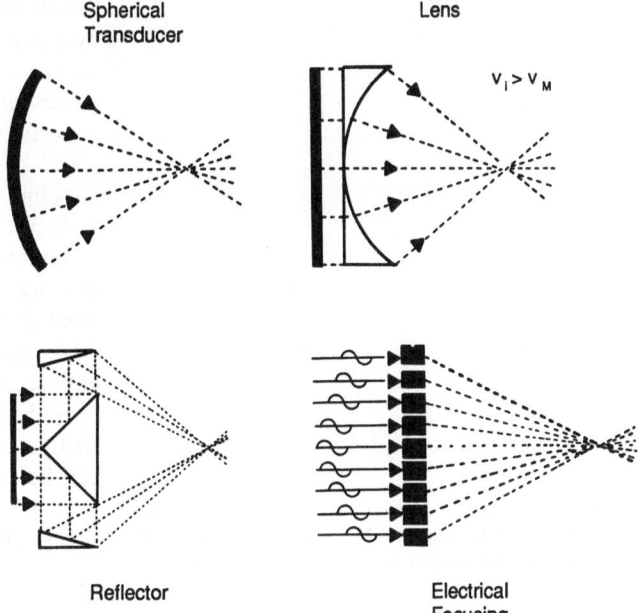

Spherical Transducer

Lens

$v_1 > v_M$

Fig. 2.10. Ultrasonic focusing systems

Reflector

Electrical Focusing

can be focused by using self-focusing radiators, lenses, or reflectors. Also focusing can be achieved by using transducer arrays that are driven with signals having the proper phase differences to obtain a common focal point (electric focusing) (Fig. 2.10). The wavelength imposes the limitation on the size of the focal region and the sharpness of the focus is determined by the ratio between the aperture of the radiator and the wavelength.

Spherically Curved Transducers. The theory of spherically curved transducers vibrating with uniform normal surface velocity has been developed by O'Neil (1949). The double integral is difficult to solve for a closed form solution and thus several numerical methods have been developed to compute the ultrasonic field from the transducers (see Sect. 2.2.2.3). Since the particle velocity and the intensity, at any point on the axis of symmetry, are directed along the axis, it is possible to write the following expression for the axial intensity distribution.

$$I(x) = I(0)\{[R/(R-x)] \sin [(\pi d^2 (R-x))/(8\lambda xR)]\}^2 \tag{2.50}$$

where I(0) is the average intensity over the radiating surface, d is the diameter of the transducer, and R is the radius of curvature. At the geometrical focus the intensity is as follows:

$$I(x) = I(0)(\pi d^2/8R\lambda)^2 \tag{2.51}$$

The acoustic pressure profile near the focal plane can be approximated by

$$p(R, \theta) = p(R, 0) \, 2 \, J_1(z)/z \tag{2.52}$$

where $z = ka \sin \theta$, $p(R, 0)$ is the acoustic pressure on the axis, and J_1 is the first order Bessel function. From the above theory, it is obvious that it is only possible to focus energy in the near-field of an equivalent plane transducer, owing to the finite size of the wavelength. (At the far-field, all of the waves coming from different points of the transducer surface are in phase in the whole beam diameter, and thus no focusing can be achieved.)

The focused acoustic field is very complex between the acoustic focus and the transducer, resembling the near-field of a plane wave transducer (Figs. 2.11, 2.12). Beyond the focus the field behaves in a similar fashion to that of the far-field of a plane transducer, except that the divergence of the beam beyond the focus is dominated by the geometrical divergence angle of the transducer. The shape of the focus is long and narrow, and these dimensions are dependent on the focusing properties of the transducer, i.e., its diameter, radius of curvature, and frequency. The geometrical focusing of a transducer is often described by an F-number, which is the ratio between the radius of curvature and the diameter of the transducer (F-number = R/d). By increasing the radius of curvature (R), the maximum intensity can be pushed deeper into the tissue, but the focal region becomes

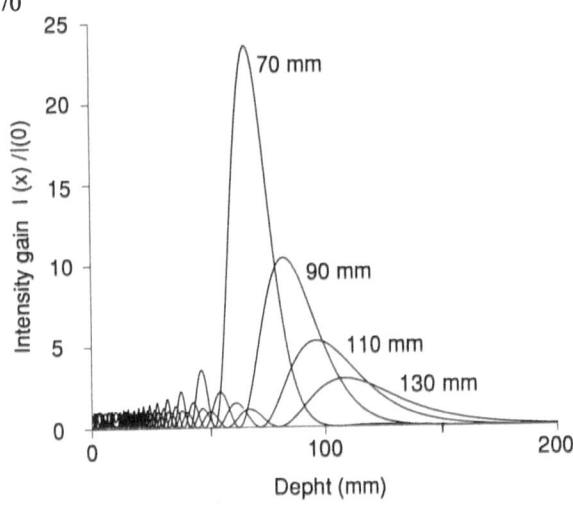

Fig. 2.11. The axial intensity distribution from a focused transducer (diameter 60 mm, frequency 1 MHz) for various values of radius of curvature calculated in tissue (attenuation 10 Npm^{-1})

longer and the peak intensity smaller. This is due to the reduced focusing effect of the transducer and the attenuation within the tissue. Thus, by increasing the radius of curvature when the diameter and frequency of the transducer are unchanged, the intensity gain at the focus will finally become less than 1 (Fig. 2.11). The maximum theoretical depth of the intensity maximum in soft tissues, as a function of frequency and diameter of the transducer, is presented in Fig. 2.13. These calculations can also be used to show the theoretical maximum depth of the focus for a certain diameter applicator, independent of the method of focusing. This is possible because a spherically curved transducer offers the best focusing, while all other methods (reflectors of equal diameter, lenses, or electric focusing) can do only as well as it and are often inferior. The graph also shows that it is possible to induce an intensity maximum at any practical depth in soft tissues with a suitable choice of the transducer parameters, as long as the beam entry is not restricted by gas or bone.

Ultrasonic Lenses. Similar to those used in optics, acoustic lenses are made of materials in which the speed of sound (v_L) is different from that in the coupling medium (v_m), causing the ultrasound beam to focus. Lenses made of solids, e.g., plastics, metals, or liquids, where the speed of sound is higher (solids) or lower (liquid) than in water, have been used. The ideal shape of a lens is planoconcave, with $v_L > v_m$, where the generating curve of the concave surface is elliptic. With small aperture angles, the spherically curved surface can be used instead of an elliptic one. In this case, the approximation for the focal length can be calculated from:

$$F = R/(1 - v_m/v_L) \tag{2.53}$$

where R = the radius of curvature of the spherically curved surface. This applies to apertures where the spherical surface does not deviate more than $\frac{1}{4}\lambda$ from the elliptical surface. This is the maximum degree of deviation because in order to add acoustic pressure at the focus, the acoustic rays should not be more than $\pi/2$ out of phase. Therefore

$$\lambda > F^2 \tan^4\theta \, \{2n^2 R[1 - (1/n) + \tan^2\theta]\}^{-1} \tag{2.54}$$

where $n = v_L/v_m$ and θ is the angle between the central axis of the lens and the line between the edge of the lens and the focal point. The aperture angle of the lens is $2x\theta$. The intensity gain (G) at the focus may be approximated by:

$$G = 0.8 \, (d^2/F\lambda)^2 \tag{2.55}$$

where d is the diameter of the lens (Fry and Dunn 1962).

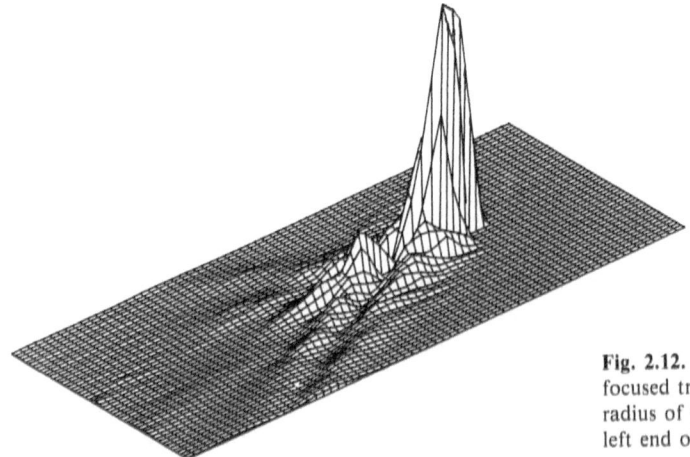

Fig. 2.12. A measured (in castor oil) field distribution from a focused transducer (frequency 1.15 MHz, diameter 75 mm, radius of curvature 70 mm). The transducer is located at the left end of the surface plot

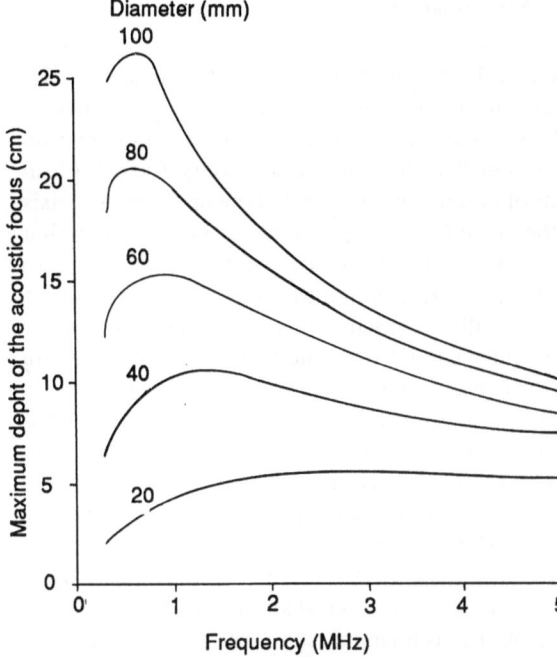

Fig. 2.13. The maximum depth of the acoustic focus of a focused transducer as a function of frequency and diameter in tissue (attenuation $10 \, \mathrm{Npm}^{-1} \, \mathrm{MHz}^{-1}$). (Hynynen et al. 1981)

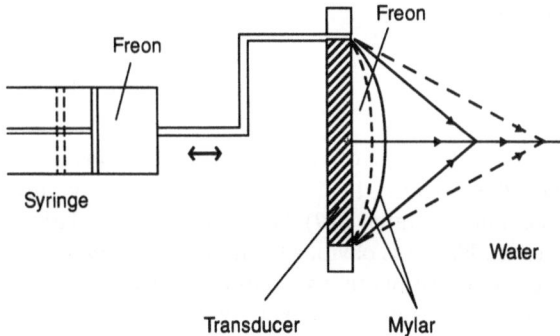

Fig. 2.14. A liquid lens with variable focal length. (Foster and Hunt 1980)

The spacing between the transducer and the lens has a significant effect on the power output of the transducer-lens system. By choosing the spacing to be $(2n-1) \, \lambda_m/4$, where n = 1, 2, 3 ... and λ_m is the wavelength in the medium, the lowest input impedance at the transducer face can be reached. Thus, a given electric field strength yields the highest transferred power with this spacing. If the spacing is 0 or $(\lambda_m/2)m$, where m = 1, 2, 3 ..., the gain is equal to or less than 1, depending on the ratio of the acoustic impedance in the lens and the medium (Fry and Dunn 1962).

The advantage of lenses over spherically curved transducers is that a desired ultrasonic field can be

produced from a single transducer by choosing the appropriate lens. By using a liquid lens and with suitable mechanical structures, a single lens can offer a wide variety of focal distances (Foster and Hunt 1980). In Fig. 2.14 a diagram of such a lens is presented. The liquid (freon), which has a low speed of sound ($900 \, \mathrm{m \, s}^{-1}$) compared with water ($1500 \, \mathrm{m \, s}^{-1}$), is enclosed between a thin, flexible membrane and the crystal, which are clamped together at the edges of the lens. The liquid space is connected to a reservoir, allowing the amount of liquid in the lens to be varied. When more liquid is pumped in, the lens becomes thicker and its radius of curvature decreases, i.e., the focal length becomes smaller. When the liquid volume is decreased the focusing effect is reduced.

Lenses have also been used to produce multiple angular foci in order to concentrate energy at the edges of a tumor, where the perfusion and thermal conduction effects are the largest. Thus, an improved temperature distribution can be obtained (Lele 1981; Beard et al. 1982). Conical lenses, which can produce long and narrow line foci, have also been proposed for use in hyperthermia (Hunt 1985; Seppi et al. 1985).

The lenses used in hyperthermia systems are usually made of plastic. The main disadvantage of the plastic lenses is their high absorption of ultrasound, resulting in temperature elevation of the lens. Since the speed of sound decreases (the focusing effect becomes smaller) and absorption increases with increasing temperature of the lens material, the shape of the field changes during high power sonication. The best plastic material for ultrasonic lenses appears to be polystyrene, in which the ultrasonic absorption is only about $5 \, \mathrm{Np \, m}^{-1} \, \mathrm{MHz}^{-1}$. Low absorption losses can also be achieved using metal lenses. However, the acoustic impedance of metals is high compared with water, and thus high losses due to acoustic mismatching can result. At present, the cost of ceramic spherically curved transducers is so low that several transducers with desired focal lengths can be obtained at a lower cost than that at which one can manufacture lenses. Thus, it is apparent that the curved transducers are the mechanical focusing method of choice for hyperthermia purposes.

Reflectors. The absorption losses caused by lenses can be avoided by using acoustic reflectors. However, the manufacturing of reflectors requires great care and is expensive. Thus, reflectors offer little advantage over focused crystals as focusing devices. However, reflectors may be useful in certain patient treatments, when the hyperthermia device cannot direct the beams to the desired location due to geometrical restrictions and patient positioning. For example, the systems where the ultrasonic transducers are immersed in a large water tank (see Sect. 2.6.4.5) do not allow the beams to be properly directed to tumors in difficult anatomical locations. In this kind of situation reflectors may prove useful.

Electric Focusing. Ultrasonic beams can be focused by using one- or two-dimensional arrays of transducers, with each element driven by signals of a specified phase so that the hemispherical waves emitted by each element (which should be small enough compared with the wavelength in order to act as a point source) are in phase at the desired focal point. The first attempts to utilize electric focusing and scanning in ultrasound hyperthermia were by Do-Huun and Hartemann (1982). They constructed a concentric ring transducer where each ring was driven with a different signal. This approach offered an acoustic focus at a desired distance on the central axis. The focus was scanned along the axis but not in any other directions. In theory such a concentric ring applicator should also be able to produce a ring focus at the focal plane (Cain and Umemura 1986). To generate an annular focal region each of the rings has to be driven with a proper phase. If the desired heating annulus has a focal length of F and radius R from the center of the focal plane, then the geometrical requirements give the phase of the i-th ring (ϕ_i) to be

$$\phi_i = -\omega_0 \, (F^2 + (R-id)^2)^{1/2}/v \qquad (2.56)$$

where ω_0 is the angular frequency of the ultrasonic wave and d is the width of the rings. Now the driving signal of the i-th ring is

$$U_i = A_i \sin(\omega_0 t - \phi_i) \qquad (2.57)$$

where A_i is the amplitude of the driving signal. According to the simulation study of Cain and Umemura (1986) such an array is capable of generating a ring focus, but it also creates a secondary focus both in front of and behind the focal plane on the central axis of the beam (Fig. 2.15). An interesting approach that avoids the secondary focuses of concentric ring is to combine electric scanning and focusing with a mechanically focused transducer (lens system or spherically curved transducer) (Cain and Umemura 1986). The principle of this system has been described in Fig. 2.17. For this device, a circular transducer has been divided into sectors, each of which has an individual set of driving circuitry. In front of the transducer there is a lens that focuses the beam. The phase difference between the driving signals of each transducer element can be set such that the pressure amplitude along the central axis is zero and the energy is focused into a ring at the focal plane. Let us now consider a sector-vortex array of radius r_1 divided into N sectors of equal size (Fig. 2.17) and with an acoustic lens of focal length F placed in front of the transducer disk. The phase ϕ_i of the i-th sector is given by

$$\phi_i = m(\theta_i + \beta(\theta_i)) \qquad (2.58)$$

where θ is the angle of rotation along the transducer plate, $\theta_i = i \, 2\pi/N$, $i = 1, 2, \ldots N$, m is the vortex mode number, and $\beta(\theta)$ is the phase modulation function. Now it is obvious from the symmetry of the system that the phase distribution produces zero intensity along its central axis when m is not zero. The

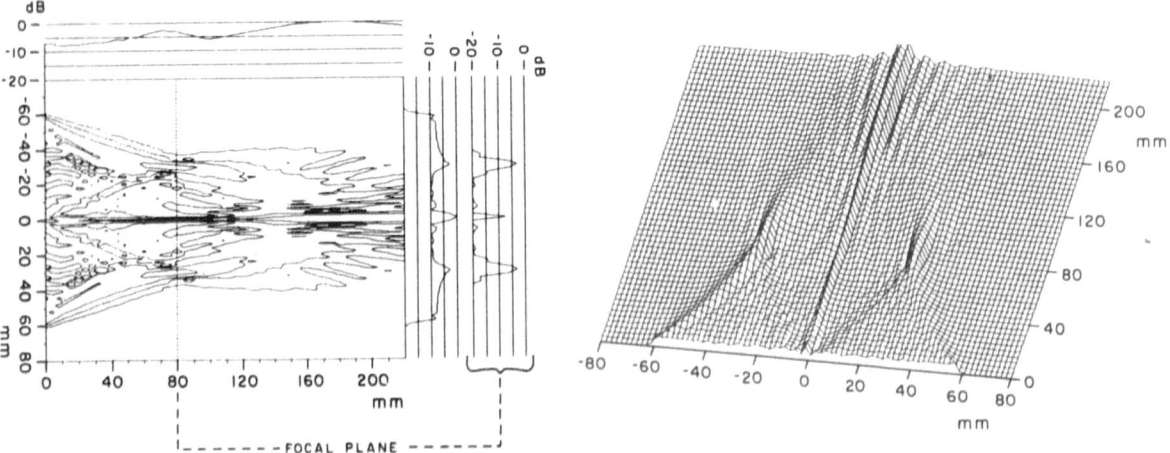

Fig. 2.15. A simulated intensity pattern produced by a concentric ring applicator. Both the intensity contours (5-dB increments) (*left*) and the surface plot (*right*) are shown. (Printed with permission; Cain and Umemura 1986; © 1986 IEEE)

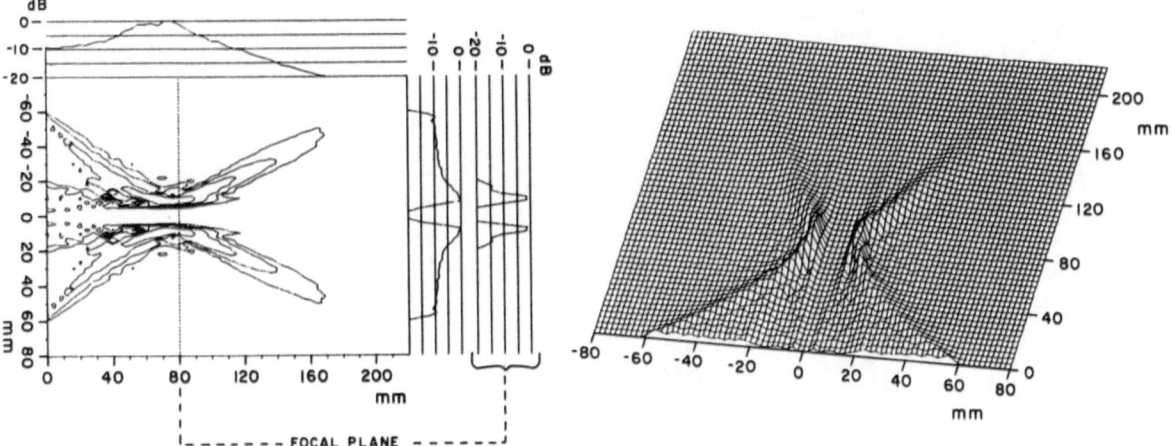

Fig. 2.16. The simulated intensity pattern produced by the sector-vortex phase array. *Left:* the contour plot (5-dB increments); *right:* the surface plot. (Printed with permission; Cain and Umemura 1986; © 1986 IEEE)

complex amplitude of the driving signal for the i-th segment is

$$A(\theta_i) = A_0 \exp \{ j [m(\theta_i + \beta(\theta_i))] - \omega_0 t \} \qquad (2.59)$$

where A_0 is a constant and $j = (-1)^{1/2}$. The field rotates around the disk at a phase velocity ω_p:

$$\omega_p = \omega_0 \{ m [1 + d\beta(\theta)/d\theta] \}^{-1} \qquad (2.60)$$

The heating pattern can be modified by varying m and $\beta(\theta)$. When m = 0 a single focus on the axis at the focal plane is formed. By increasing m the radius of the focal ring increases. The shape of the focal ring can be modified with $\beta(\theta)$. A circular focus is produced when $\beta(\theta) = 0$ (Fig. 2.16). If $\beta(\theta)$ is not zero an elongated heating field can be generated.

The field from the sector-vortex array also can be focused with means other than the lens presented in Fig. 2.17. A spherically curved transducer can be divided into sectors and driven with the previously mentioned phase difference, producing a similar result. The focusing can be done electrically by dividing the sector array into concentric rings and then driving each element with a signal that would give first a focus and then a rotating field. Such a combined array offers a lot of flexibility.

The most flexible, although expensive way to utilize electric focusing is to use a two-dimensional array of small transducers, each of which has a separate amplitude and phase control (i.e., each line has its own amplifier, phase shifter, and attenuator). Let us consider a two-dimensional (x-y) array of transducers (element dimensions Δ x and Δ y). In order to produce a focus at the distance of F at the central axis of the array, all of the waves coming from the different

elements have to be at the same phase in the focus. This means that the phase of the i, k-th element $\phi_{i,k}$ (i and k indicate the number of the element in x- and y-directions from the center) is

$$\phi_{i,k} = -\omega_0 [F^2 + (i\Delta x)^2 + (k\Delta y)^2]^{1/2}/v \qquad (2.61)$$

where ω_0 is the angular frequency and v is the propagation speed of the wave. This formula can be easily adapted for any other location of the focal point. This shows that the focal point can be electrically scanned along a preselected pattern by controlling the phase shift of the signal driving each transducer. Electric focusing and scanning offers a flexible and fast way to control the location and movement of the focal point. However, it suffers from the need for a complex electronic driving circuitry, with two-dimensional arrays.

In order to reduce and simplify the required electronics several methods have been developed. Ocheltree et al. (1984) reported a stacked array approach where only three rows of transducers are driven at one time and then the excitation is sequentially switched to include two from the previous group and a row next to them and so on until the whole array has been used. This approach will considerably reduce the number of separate driving lines. An alternative is to construct the array from long narrow transducers the thickness of which increases gradually from one end to the other. Driving a single transducer with a certain frequency would generate the strongest ultrasonic field at the location the thickness of which is one-half of the wavelength. Thus, by scanning the driving frequency the active location in the transducer can be moved to a desired location. By constructing an array of these tapered transducers, three-dimensional scan-

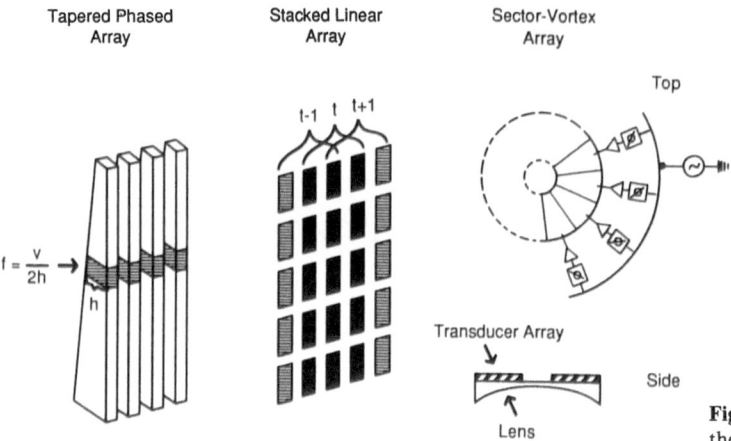

Fig. 2.17. Three different methods to reduce the electronics required during electrical scanning and focusing

ning can be obtained (Benkeser et al. 1985) (Fig. 2.17).

Another method of utilizing both mechanical and electric focusing was described by Ebbini et al. (1988). Using simulations they showed that the ultrasonic field from a cylindrical sector array of transducers could be electrically focused and scanned to induce therapeutic temperature elevation in deep tumors. These array examples have shown how flexible electric focusing is, and it appears that this approach has a significant potential in the future ultrasound hyperthermia treatments.

2.2.2.3 Acoustic Field Calculations

In this section, a more detailed description of some of the techniques used to calculate the acoustic field will be discussed (see for example Swindell 1986). Let us take a radiating surface with area A. If the velocity normal of the surface (u_a) is uniform and proportional to $e^{j\omega t}$, where t is time and $j = (-1)^{1/2}$, then the resulting velocity potential (Ψ_p) at a field point p can be approximated by the Rayleigh-Sommerfield diffraction integral:

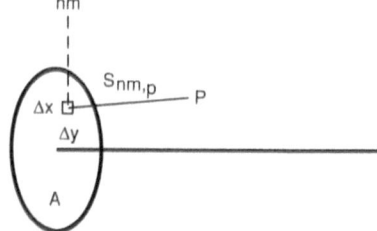

Fig. 2.18. The summation method for the calculation of the ultrasonic field

$$\bar{\Psi}_p = \iint_A \frac{u_a}{2\pi s} e^{-jks} dA \qquad (2.62)$$

where s is the distance from the source point dA to the field point p. Now, the particle velocity (u) and the acoustic pressure at the field point are given by

$$\bar{u}_p = -\text{grad } \bar{\Psi}_p \qquad (2.63)$$

and

$$p = \varrho \partial \bar{\Psi}/\partial t = j\omega\varrho \bar{\Psi} \qquad (2.64)$$

where ϱ is the density of the medium.

Equation 2.62 is difficult to solve in its closed form. However, there are several numerical approximations that can be used to obtain the acoustic velocity potential distribution generated by the transducer. The first technique used a two-dimensional numerical integration over the whole surface of the transducer. In this technique, the transducer surface area, A, is divided into small sources (small enough to be considered as point sources) and then all of the waves coming from these elements to point p are summed (Fig. 2.18). For a flat transducer this can be expressed as

$$\bar{\Psi}_p = \frac{\Delta x \Delta y}{2\pi} \sum_{n=1}^{N} \sum_{m=1}^{M} \frac{u_{nm} e^{-jkS_{nm,p}}}{S_{nm,p}} \qquad (2.65)$$

where $S_{nm,p}$ is the distance from the point source nm to the field point p and $u_{nm} = u_0 (nm) e^{j\omega t}$ is the normal velocity of the element nm. Since the element size $\Delta x \Delta y$ has to be small compared with the wavelength, the calculations become very time consuming (Zemanek 1971).

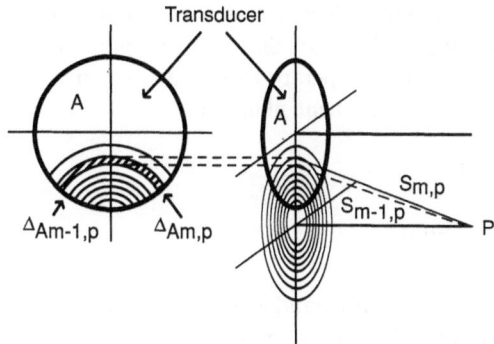

Transducer

$S_{m,p}$
$S_{m-1,p}$
P
$\Delta A_{m-1,p}$ $\Delta A_{m,p}$

Fig. 2.19. The technique to reduce the double integral to an one-dimensional integral. (Printed with permission; Swindell et al. 1982; Roemer et al. 1984; © 1984 IEEE)

There are several techniques which can be used to reduce the double integral into an one-dimensional integral (Madsen et al. 1981; Swindell et al. 1982). If the normal velocity u_a is constant over the whole surface area, then an estimate of the velocity potential at field point p is given by:

$$\Psi_p = \frac{u_a}{2\pi} \sum_{m=1}^{M} \frac{\Delta A_{m,p}}{S_{m,p}} e^{-jkS_{m,p}} \qquad (2.66)$$

where $A_{m,p}$ is the element area over which the wavelet with radius $S_{m,p}$ is constant (Fig. 2.19). This element area can be found by considering a spherical surface with radius $S_{m,p}$ centered at point p. This sphere intersects the transducer surface. The element area is located about the intersection arc. Now, by choosing the strip narrow enough, a desired computational accuracy can be achieved with both planar and spherically curved transducers. This model has been compared with the field generated by a real focused ultrasound source and has been found to give a reasonable approximation of the field (Madsen et al. 1981).

Another technique used to speed up the calculations is the Fourier method, where A will be calculated as a Fourier transform. The equation can then be solved using fast Fourier transform algorithms (Swindell 1986).

When the ultrasound beam penetrates into the tissue, it will be attenuated due to the absorption and scattering of the energy. This has to be taken into account when the ultrasonic fields are calculated for induction of hyperthermia. In an uniform tissue, for plane waves or weakly focused fields, the velocity potential in the point p can be approximated as

$$\Psi_p = \frac{\Delta x \Delta y}{2\pi} \cdot e^{-\mu z} \sum_{m=1}^{M} \sum_{n=1}^{N} \frac{u_{nm}}{S_{nm,p}} e^{-jkS_{nm,p}} \qquad (2.67)$$

where z is the depth the beam has penetrated into the tissue and μ is the amplitude attenuation coefficient of the tissue. However, for strongly focused fields with high frequencies, the attenuation should be taken into account when the contribution of each area element dA is calculated for the point p. Similarly, in layered media the attenuation contributed by each tissue layer should be taken into account. When the beam is penetrating through tissue layers with different attenuation coefficients the velocity potential at point p can be written as follows:

$$\Psi_p = \frac{\Delta x \Delta y}{2\pi} \sum_{m=1}^{M} \sum_{n=1}^{N} \frac{u_{nm}}{S_{nm,p}} e^{-jkS_{nm,p}} e^{-\sum \mu_i \cdot S_{nm,p,i}} \qquad (2.68)$$

where μ_i is the attenuation in a tissue layer i and $S_{nm,p,i}$ is the portion $S_{nm,p}$ that penetrates through the layer i.

The effect of various soft tissue interfaces on the attenuation or divergence of an ultrasonic beam has generally been assumed to be small. However, bone or gas interfaces reflect ultrasonic energy strongly and thus in some locations there is also a reflected wave component in addition to the directly incident wave. This wave should be taken into account when the velocity potential at a field point is calculated.

Part of the energy is also scattered in all directions into the tissues. This energy is lost from the beam and does not contribute to the temperature elevation at that point. However, the scattered wave from other tissue volumes contributes to the wave in each location and thus the scattered waves should be taken into account when the absorbed energy is estimated. Generally, in the ultrasound hyperthermia simulations, all of the attenuated energy has been assumed to be absorbed locally and the scattering has been ignored. Clearly, the scattered energy will be absorbed sooner or later, and during ultrasound hyperthermia most of the scattered waves are probably absorbed in the treatment volume. However, at present there is no experimental evidence as to how important the scattering is and how it should be taken into account in the calculations. It is also clear that accurate in vivo field measurements should be performed to make possible the development of precise ultrasonic field calculation programs.

When the velocity potential is known at a point p, the amount of absorbed energy $\langle q_p \rangle$ can be calculated by using Eqs. 2.34 and 2.64 to yield the following:

$$\langle q_p \rangle = \alpha \, \omega^2 \, \varrho^2 \, |\Psi|^2 / Z \qquad (2.69)$$

where α is the amplitude absorption coefficient and

$\omega = 2\pi \mathrm{v}/\lambda$ is the angular frequency of the sound wave.

From the above numerical methods, the absorbed energy can be calculated throughout a modeled tissue volume. With this knowledge, the resulting temperature distributions can be computed for various treatment situations. These temperature distributions in tissue are strongly dependent on the blood perfusion rate and thermal conductivity. Therefore, in order to obtain the temperature distributions from the absorbed power values, the bio-heat-transfer-equation (Pennes 1948) has been used in the simulation studies (for example Hynynen et al. 1981; Swindell et al. 1982; Roemer et al. 1984; Moros et al. 1988, 1989; Cain and Umemura 1986).

2.3 Acoustic Properties of Tissues

In order for a physicist to plan ultrasound hyperthermia treatment, it is essential to know the ultrasonic properties of tissues so that the absorbed power density in the treatment volume can be estimated. For instance, the ultrasonic velocity determines the field shape and the amount of reflected energy at the tissue interfaces (because the acoustic impedance = velocity ×density). The attenuation and absorption control the magnitude of absorbed energy, and their ratio determines how much energy is scattered away from the beam. The nonlinear dependence of the absorbed energy on the intensity, as well as the threshold intensity for the appearance of direct mechanical damage (transient cavitation), has to be known in order to avoid undesirable effects. Thus, precise knowledge of the absolute values of the acoustic properties of tissues and their temperature dependence is necessary for successful treatment planning.

2.3.1 Velocity

The velocity of ultrasound is not frequency dependent and has a similar average magnitude of 1550 m s^{-1} in all soft tissues (excluding lung). The velocity in fatty tissues is less than that in other soft tissues, being about 1480 m s^{-1}, while in the lung the air spaces reduce the velocity to about 600 m s^{-1}. The highest values, between 1800 and 3700 m s^{-1}, have been measured in bones. In various soft tissues the velocity appears to increase gradually as a function of the temperature. The slope of the increase has been found to be between 0.04% and 0.08% K^{-1} (Fig. 2.20). In fatty tissues, the speed of ultrasound decreases as the temperature increases. According to Johnson et al. (1977), the velocity in fat starts to increase slowly at temperatures higher than the normal body temperature, whereas Bamber and Hill (1979) found that the velocity continued to decrease. However, this temperature dependence of the velocity of ultrasound is not important from the point of view of hyperthermia, due to its very small effect on the field shape.

2.3.2 Absorption

In an idealistic, pure elastic medium, the energy in an ultrasonic field is in either kinetic or potential form, and the pressure wave is in phase with the particle velocity. In a real medium there are also viscous forces between the moving particles, which cause a lag between the particle pressure and velocity (or change in density). Therefore, an energy loss during each cycle will result. The theoretical absorption coefficient in a viscoelastic medium is determined by Stokes-Kirchoff's classical absorption theory:

$$\alpha_{\mathrm{v}} = \frac{\omega^2}{2\varrho_0 \mathrm{v}^3} \left[\frac{4}{3} \eta_{\mathrm{s}} + \eta_{\mathrm{b}} \right] \qquad (2.70)$$

where η_{s} and η_{b} are the shear and bulk viscosity coefficients, respectively. As can be seen from Eq. 2.70, the absorption in a viscoelastic medium should depend on the square of the frequency (f^2). This is true in many liquids but not in tissues, where the absorp-

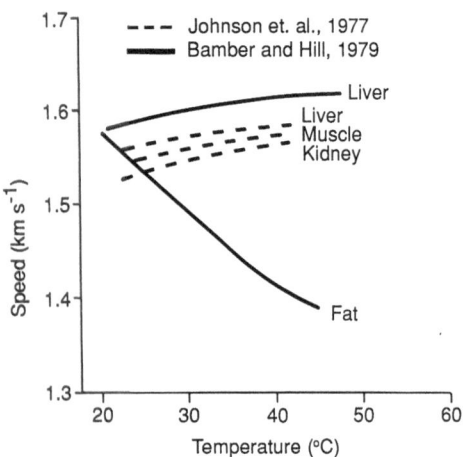

Fig. 2.20. The speed of sound as a function of temperature in various soft tissues

tion has been shown to increase almost linearly as a function of frequency. Also, the measured absorption coefficient values are significantly larger $(3-10 \, \text{Npm}^{-1}$ at 1 MHz) than those estimated based on the classic absorption theory (around $0.1 \, \text{Npm}^{-1}$ at 1 MHz). Thus, there must also be other absorption mechanisms in tissues in addition to the viscous one. During the compressive part of the cycle, energy is stored in the medium in a number of forms, such as lattice vibrational energy, molecular vibrational energy, and translational energy. During the expansion part of the cycle, this stored energy is returned to the wave and the medium temperature returns to the original level. However, in tissue the increased kinetic energy of the molecules is not in balance with the environment, and the system tries to redistribute the energy. The transfer of energy takes time and thus, during the decompression cycle, energy will return out of phase to the wave and absorption results. In addition, a portion of the stored energy remains stored in various forms within the medium. This mechanism of energy absorption is called *relaxation*.

In order to illustrate the relaxation process let us consider a step function pressure change (p) in a medium (Fig. 2.21). The medium response is not immediate, and it takes a certain amount of time before the response reaches the new level. The response can be characterized by a time constant t_r, which is the time required to reach 50% of the complete response. Now, the relaxation frequency f_r can be calculated as

$$f_r = 1.44/(2\pi t_r) \qquad (2.71)$$

The absorption coefficient increases with increasing frequency until it reaches a maximum value when $f = f_r$. At higher frequencies there is less time to redistribute the energy in the medium between the compression and rarefactions and less energy is absorbed (Fig. 2.22). At the same time the speed of sound in the medium increases and reaches its maximum, and remains constant at this level for frequencies above the relaxation frequency. This phenomenon is called *dispersion*. Dispersion can be attributed to the reduction of energy redistribution during the cycle, and thus the medium behaves like it has an increased stiffness at higher frequencies.

The absorption in biological tissues cannot be explained by a single relaxation process; a spectrum of relaxation frequencies (f_{ri}) is required. Thus, the absorption may be written as:

$$\frac{\alpha}{f^2} = A_c + \sum_i \frac{B_i}{1+(f/f_{ri})^2} \qquad (2.72)$$

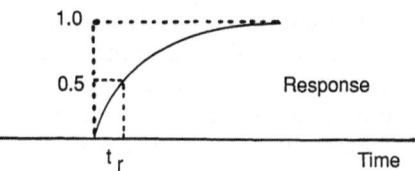

Fig. 2.21. A pressure change (*top*) and the system response as a function of time (*bottom*)

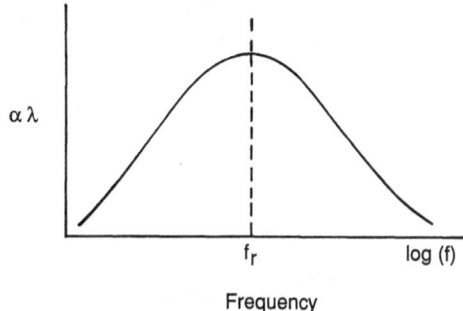

Fig. 2.22. The absorbed energy per cycle as a function of the frequency when the absorption is caused by one relaxation process with the relaxation frequency f_r

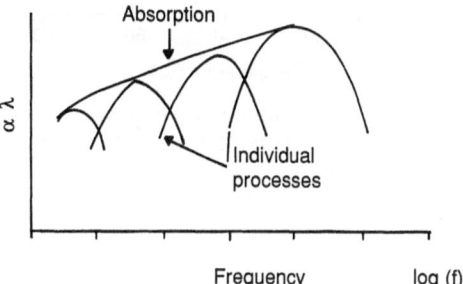

Fig. 2.23. The absorbed energy per cycle as a function of frequency when multiple relaxation processes are involved

where A_c represents the classical absorption term and B_i is a constant for a relaxation process i. The effect of multiple relaxation frequencies is shown in Fig. 2.23. The ultrasonic absorption mechanism in tissues has been reviewed in detail by Dunn (1976), Wells (1977), and Mortimer (1982).

The measured absorption coefficient of tissue appears to increase as a function of frequency in accordance with the following relations:

$$\alpha = \alpha_0 \, (f)^m \qquad\qquad (2.73)$$

where α_0 is the absorption coefficient/MHz and f is the frequency in MHz. α_0 and m are dependent on the tissue type and m has been found experimentally to be between 1 and 1.3 (Goss et al. 1979; Lyons and Parker 1988). Goss et al. (1979) measured the absorption coefficients for various tissues in cat, mouse, pig, and cow and found little difference among the species studied. However, there were more variations among the absorption coefficients of the different tissues. Their results, together with other values from the literature, are summarized in Table 2.1. Generally, the absorption coefficient in soft tissues appears to be around 3 $\mathrm{Npm^{-1}\,MHz^{-1}}$, excluding the tendon and testis, which have absorption coefficients of 14 and 1.5 $\mathrm{Npm^{-1}\,MHz^{-1}}$, respectively.

The absorption coefficient in central nervous tissues has been found to be temperature dependent (Dunn and Brady 1973). At 1 MHz the absorption increases

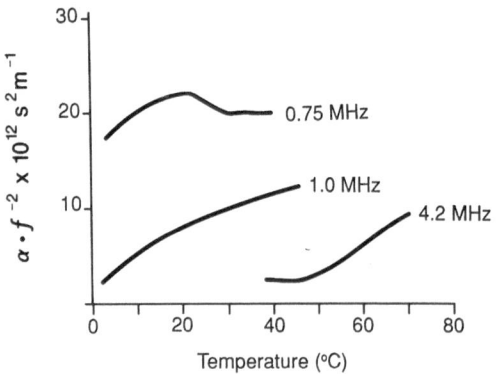

Fig. 2.24. The effect of temperature on ultrasonic absorption in mammalian central nervous system (Dunn and Brady 1973)

with increasing temperature up to 45 °C, whereas at higher or lower frequencies it appears to be almost independent of temperature in the hyperthermic range (37°–45 °C) (Fig. 2.24). More data for different tissues are needed before any firm conclusions can be drawn on the importance of the absorption changes during ultrasound hyperthermia treatments.

2.3.3 Attenuation

Ultrasonic attenuation in tissues is a sum of the losses due to absorption and scattering. In the scattering process the elastic discontinuities within the tissue scatter the ultrasound beam away from its original direction of propagation. In experimental studies, the attenuation has been found to follow a frequency dependence similar to that of the absorption. However, the attenuation values have been found to be larger than those of absorption. Goss et al. (1979) compared the attenuation and absorption values in various tissues for similar experimental situations and found the attenuation-absorption ratio to be about 3. They explained that part of this difference was due to errors in attenuation measurement. More recently, this explanation received some experimental support, when Carstensen et al. (1981) and Lyons and Parker (1988) experimentally obtained almost the same values for attenuation and absorption of in vitro liver. The attenuation values determined using different techniques were found to vary, but overall these results indicate that the reported attenuation values are too high. The highest attenuation values have been measured for lung (430–480 $\mathrm{Npm^{-1}}$ at 1 MHz) and bone (150–350 $\mathrm{Npm^{-1}}$ at 1 MHz). The soft tissue

Table 2.1. Acoustic properties of mammalian tissues at a temperature of 37 °C and a frequency of 1 MHz (data from Goss et al. 1978, 1979, 1980; Chivers and Parry 1978; Lyons and Parker 1988; Wells 1977)

Tissues	Velocity (m s^{-1})	Attenuation (Npm^{-1})	Density (kg m^{-3})	Acoustic impedance (10^6 kg m^{-2} s^{-1})	Absorption (Npm^{-1})
Bone	1500–3700	150 – 350	1380–1810	3.75 – 7.38	–
Brain	1516–1575	4 – 29	1030	1.56 – 1.62	1.2[d]–6.4[e]
Fat	1400–1490	5 – 9	921	1.29 – 1.37	–
Kidney	1564–1640	3 – 10	1040	1.62 – 1.71	3.3
Liver	1540–1640	3.2– 18	1060	1.70 – 1.74	2.3 – 3.2
Lung	470–658	430 – 480	400	0.188–0.263	7
Muscle	1508–1630	4.4– 15[b]	1070–1270	1.61 – 2.07	2 – 11
Skin	1498[a]	14 – 66[c]	1200	1.80	–
Tendon	1750	30 – 70	–	–	14
Testis	1595	1.5– 3.8	–	–	1.5

Temperature: [a] 23 °C, [b] 40 °C, [c] not reported
Brain: [d] grey matter, [e] white matter

values are generally about 10 Npm^{-1} MHz^{-1}. Fatty tissue has lower values, between 5 and 9 Npm^{-1} MHz^{-1} (Table 2.1). The ultrasonic attenuation in some neoplastic tissues has been found to be higher than in normal host tissues. Kikuchi et al. (1957) reported that the attenuation in brain tumors is higher than in the normal brain at frequencies between 0.5 and 5 MHz. For example, the attenuations in a meningioma and normal brain tissue were 60 Npm^{-1} and 20 Npm^{-1} at 1 MHz, respectively. Similar differences have also been found between malignant breast tumors and normal breast tissue by Calderon et al. (1976).

The relationship between ultrasonic attenuation and temperature depends on the frequency. Some of the data reported by Bamber and Hill (1979) for bovine tissue are presented in Fig. 2.25. The temperature dependence of ultrasonic attenuation appears to increase with increasing frequency. At the frequency of 1 MHz, the attenuation remains constant or changes only slightly up to body temperature, above which it increases gradually. In many of the tissues tested, the attenuation values appeared to remain practically constant within the temperature range used in hyperthermia, at the frequency of 1 MHz. The denaturation of the tissue proteins at the higher temperatures increases the attenuation coefficient strongly in brain tissues (Robinson and Lele 1972).

2.3.4 Characteristic Acoustic Impedance

As was shown in the theoretical discussion of Sect. 2.2, the acoustic impedance of a tissue is the product of the velocity of the wave and the density of the medium. Generally, most of the soft tissues have an impedance roughly equal to that of water, having a density around 1000 kg m^{-3} and an acoustic impedance of 1.6×10^6 kg m^{-2} s^{-1}. Fat has a slightly lower impedance value of 1.35×10^6 kg m^{-2} s^{-1} due to its lower density and lower speed of sound. Bone and lung have impedances significantly higher and lower, respectively. In practice, these impedance differences mean that the ultrasound beam suffers little reflection loss while penetrating from one soft tissue to another, but a significant amount of energy is reflected at soft tissue-bone interface. At a tissue-gas interface, all the energy is reflected back into the tissue. The reflections of sound from these interfaces must be taken into account when treatments are planned, otherwise unexpected hot spots may occur.

2.3.5 Shear Wave Properties

At interfaces between different tissues, the longitudinal ultrasonic waves may be converted into shear waves. The attenuation of shear waves is much higher than that of longitudinal waves, being about 15×10^3 Npm^{-1} at 1 MHz (Frizzell and Carstensen 1976; Madsen et al. 1983). This mode conversion is important at the interfaces between soft tissues and bones. The magnitude of shear wave generation is a function of the angle of incidence, and reaches its maximum between 45° and 60° (Chan et al. 1974) (Fig. 2.26). It is not known whether shear wave generation is an important factor in the wave attenuation at other tissue interfaces.

2.3.6 Nonlinear Propagation Parameter

As discussed in Sect. 2.2.1.5, the propagation of a high intensity ultrasound beam is not linear. The nonlinearity of the wave propagation causes wave distortion and the formation of higher harmonics, which are attenuated more rapidly than the fundamental frequency, resulting in an increased energy absorption and a higher temperature elevation than expected from linear propagation. It may even cause acoustic saturation, which limits the ultrasonic energy transferred over a known distance, from the transducer into the biological tissue. The use of focused ultrasound makes it possible to obtain larger ultrasound intensities in the target volume, and thus nonlinear propagation may be utilized to give higher absorbed power densities in the tumor than could be obtained with just linear propagation. This may also

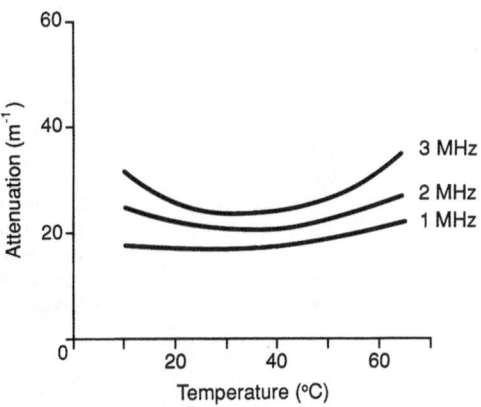

Fig. 2.25. The effect of temperature on ultrasonic attenuation in bovine liver (Bamber and Hill 1979)

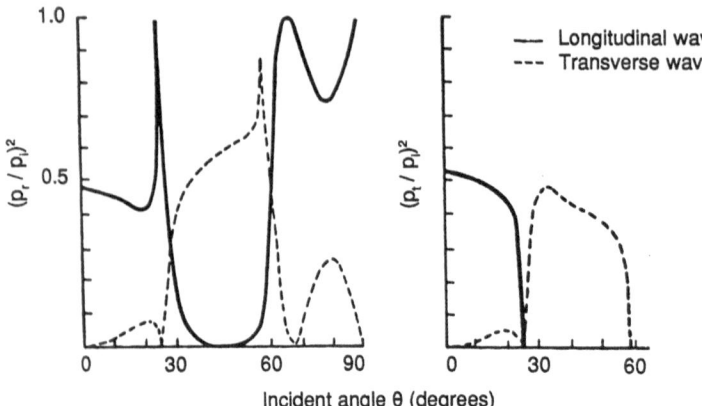

Fig. 2.26. Mode conversion (= shear wave generation) at the water-plexiglas interface (Mayer 1965)

reduce the amount of energy propagating beyond the focal and target volume, thus decreasing the possibility of undesired hot spots at a bone surface beyond the tumor.

The measurement techniques to determine the nonlinear propagation parameter B/A have been reviewed in detail by Bjorno (1986) and Carstensen and Muir (1986) and some measured values for various tissues are given in Table 2.2. It appears that the highest values of B/A, up to 11, have been found for pig fat. According to Apfel (1986), the rise in tissue fat content increases the value of B/A. In other soft tissues, the measured values of the nonlinear propagation parameter have in general been between 7 and 8, thus being lower than those in fat but higher than those in water (about 5.4). The B/A value also appears to increase with rising temperature in water and in human multiple myeloma. However, its temperature dependence in fat is not clear.

2.4 Biological Effects of Ultrasound

The basic mechanisms of the biological effects of ultrasound will be discussed in this section. As was pointed out in Sect. 2.2, an ultrasound field causes particle motion which results in mechanical stress and strain. Sometimes these mechanical forces can cause direct changes in the biological system. The *mechanical interactions* between ultrasound and tissue include radiation force and pressure, radiation torque, and streaming (shearing stress).

At high intensity levels the biological effects are sometimes associated with the formation of small gas bubbles or the oscillation of those small bubbles already present. This type of interaction is called *cavitation*, and it can cause complete destruction of the tissues located next to the gas bubble.

In an absorbing medium, the ultrasonic energy is continuously absorbed and converted into a temperature

Table 2.2. Values of the nonlinearity parameter B/A for biological tissues

Tissue	Temperature (°C)	B/A	Reference
Pig fat		11.0 – 11.3	Law et al. 1985
Human breast fat	20	9.2	Sehgal et al. 1984
	30	9.9	
	37	9.6	
Pig muscle	30	7.5 – 8.1	Law et al. 1985
Beef liver	23	7.3 – 8.1	Law et al. 1985
Beef brain	30	7.6	Law et al. 1985
Beef heart	30	6.8 – 7.4	Law et al. 1985
Human multiple myeloma	22	5.6	Sehgal et al. 1984
	30	5.8	
	37	6.2	
Water	20	5.0	Sehgal et al. 1986
	30	5.2	
	37	5.4	

elevation within the medium. If the temperature elevation is large enough and is maintained for a certain period, the exposure will result in tissue damage. This *thermal effect* is similar to that obtained using other heating systems with equal thermal exposure.

Most of the studies that have investigated the biological effects of ultrasound have been executed at intensities and exposures that are more related to diagnostic ultrasound than hyperthermia (see reviews: Williams 1983; Nyborg and Ziskin 1985; NCRP 1983; Carstensen 1987). So far, the best indication of the biological effects of therapeutic ultrasound comes from the studies in which ultrasound hyperthermia treatments have been compared with other methods of inducing similar temperature elevations in a controlled environment. These hyperthermia treatments have produced equal responses for cases where the temperature distributions were similar (Hahn 1982). This indicates that the thermal effects are dominant over the mechanical and cavitational effects, at least for the intensities and frequencies studied, and therefore ultrasound is as safe as these other modalities. However, knowledge of the importance of the nonthermal effects of ultrasound during hyperthermia is very limited. More in vivo studies are needed to clarify the role of these interactions in the treatment of cancer, especially when ultrasound hyperthermia is combined with radiation and/or chemotherapy.

2.4.1 Thermal Effects

The thermal effects produced by ultrasound are utilized in hyperthermia as a cancer therapy. Similar to any of the other types of hyperthermia treatment, its effectiveness depends on the temperature achieved and the length of the exposure. In addition, the tissue type, some physiological factors (pH and O_2), and other factors have a strong influence on the induced biological effects. A certain intensity or power of the ultrasonic field does not necessarily induce a known temperature elevation, and this is also true of microwave power. The temperature elevation in the tissue depends on the absorption and attenuation coefficients of the tissue, the size and shape of the heated region (thermal conduction effects), and also very strongly on the local blood perfusion rate. In order to illustrate briefly the temperature-time relationship of some biological effects, the threshold for a coagulation necrosis induced with focused ultrasound beams is plotted in Fig. 2.27 (Lele 1977).

2.4.2 Mechanical Effects

To illustrate the direct mechanical forces acting on the particles in the biological medium, consider a situation where a 1-MHz beam with an intensity of $100\ W\ cm^{-2}$ is scanned over a tissue volume. Note that with focused and scanned beams used during hyperthermia treatments the intensities can be even larger. Now, the particle displacement, maximum velocity, and acceleration are $0.18\ \mu m$, $1.15\ m\ s^{-1}$, and 7.4×10^5 gravity, respectively. The maximum displacement occurs over half of the wavelength, which is about $0.75\ mm$ at this frequency. Thus, the stress caused by the particle displacement is not large, and the direct rupture of the cell membranes is unlikely. The situation is quite different if shock waves have been formed, where the particles are under much larger mechanical forces.

The radiation force is a steady force exerted on objects in ultrasonic fields and is often considered a second order mechanical effect. The radiation force in a standing wave field has been found to cause red blood cells, in a vessel of a chick embryo, to align in bands with a spacing of one-half of the sonic wavelength (Dyson et al. 1974). In a traveling wave the radiation force probably has a much smaller effect.

Another secondary force that is closely related to the radiation force is radiation torque, which tends to produce rotary motion. Spinning of intracellular organs in ultrasonic fields has been reported by several investigators; for example, Martin et al. (1983) studied this effect in a fish tail in vivo. Radiation torque causes motion on the cellular level when the cell walls, intracellular structures, or gas-filled spaces cause inhomogeneities in the sound field allowing for

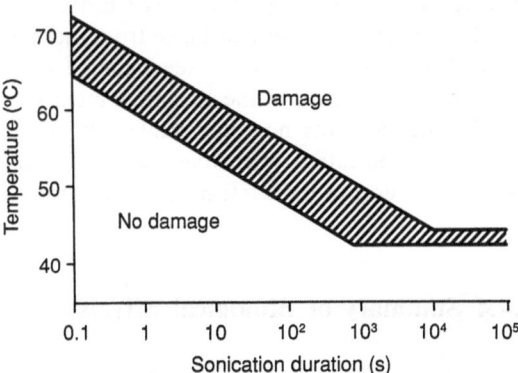

Fig. 2.27. The tissue damage threshold temperature as a function of sonication time. The threshold temperature induced with other heating methods also falls inside the *shaded area*. (Lele 1977)

an imbalance of the macroscopic forces. If spinning of the intracellular structures happens during cell division it can cause abnormal daughter cells (Dyer 1972). The acoustic torque can lead to steady circular flow called acoustic streaming. When the streaming is caused by a vibrating wire or a gas bubble, strong shear forces can be generated which may lead to the destruction of the cells. For example, a small gas bubble oscillating in a blood vessel at 80 mW cm^{-2} and in a 1-MHz ultrasonic field in vivo caused the aggregation of platelets (Frizzell et al. 1986). Thus, the shear stresses induced by microstreaming can be strong enough to cause severe tissue damage. In many situations where a bioeffect produced by ultrasound cannot be explained by increased temperature effects or cavitation these second order mechanical quantities have been used to explain the result.

2.4.3 Cavitation

Acoustic cavitation can be defined as the interaction of a sound field with the microscopic gas bodies in an exposed medium. In order for cavitation to occur, the presence of small gaseous nuclei, which probably exist in mammalian tissues, is required. When a medium that can host such cavitation centers is sonicated the bubbles start to expand and contract in a fashion that is inversely proportional to the acoustic pressure. The pulsation amplitude reaches its maximum value near a characteristic frequency corresponding to the volume resonance of the bubbles. The resonance size of the bubbles in a 1-MHz frequency ultrasound field is only 3.5 µm and decreases with increasing frequency. Smaller bubbles tend to grow toward the resonance size by rectified diffusion, a process in which more gas is diffused into the bubble from the surrounding medium during the expansion state than is returned during the compression phase. Gas bubbles larger than the resonance size do not interact in the cavitation process. When a bubble is sonicated at its resonance frequency it may intercept and erradiate energy, thereby absorbing much more acoustic power than would pass through normal tissue of its geometrical cross-section. This type of bubble oscillation is called *stable cavitation* and can cause microstreaming of the fluids around the bubble. As a consequence the previously mentioned highly localized shear stresses may lead to severe cell damage. There is experimental evidence of the generation of these microbubbles in a 0.75-MHz ultrasonic field at intensities as low as 0.68 W cm^{-2}, with the number of bubbles increasing at higher intensities (Ter Haar et

al. 1982). Lele (1977) reported similar stable cavitation with diagnostic intensities at 2.7 MHz, but could not observe any tissue damage due to cavitation when the tissue samples were histologically examined. In any case, if stable cavitation occurs during ultrasound hyperthermia it can significantly increase the power attenuation and thus cause unexpected temperature elevations in tissue layers through which the tumor is sonicated. At present, there is no experimental evidence regarding the influence of stable cavitation on either the tissue damage or the propagation of ultrasound beams during hyperthermia treatments.

At high enough acoustic pressures, the bubble oscillations become highly nonlinear and the bubbles may expand and collapse violently. The transition from the stable cavitation to this *transient cavitation* occurs with a small increase in pressure and thus the pressure value for the onset of this phenomenon is called the threshold for transient cavitation. A shock wave is emitted during the collapse of the bubble, causing mechanical stress. During the collapse, the acoustic pressures can be as high as several thousand atmospheres and the temperatures can reach several thousand degresses of Kelvin. This results in the formation of free radicals ($-H$ and $-OH$) which are chemically active. Despite its severe nature, the effects of transient cavitation are very localized, extending only a cubic micrometer or less. The threshold pressure for transient cavitation increases as a function of frequency, and of a decrease in duty cycle and pulse length (Hill 1972). In tissues with continuous wave sonication the threshold intensity has been found to be 75 W cm^{-2} at 368 kHz (Sommer and Pounds 1982) and 1450 W cm^{-2} at 2.7 MHz (Lele 1977). The latest results indicate that cavitation may be responsible for the tissue damage produced at the intensity of 289 W cm^{-2} at 1 MHz in vivo (Frizzell et al. 1983). According to Lele (1977), transient cavitation was associated with hemorrhage and tissue disintegration and the damage was distinguishable from thermal lesions. The location of the tissue damage did not always occur at the site of the maximum intensity and thus its location was not predictable. Therefore, at present the sonication intensities should be kept at levels which do not induce transient cavitation.

2.4.4 Summary of Biological Effects

Up until now, hyperthermia experiments that have used ultrasound to induce the temperature elevation have not shown the sonication to produce any enhanced cytotoxic effects. However, the temperature distributions induced in vivo are not uniform and it is extremely difficult to induce equal thermal exposure, even with the same method. Thus, the nonthermal interactions may have been shadowed by variations in the thermal exposure. In addition, the nonthermal effects may become significant only when the hyperthermia treatment is combined with radiation or chemotherapy. There are some in vitro cell studies (Ter Haar et al. 1980; Li et al. 1977) which indicate that the cell-killing effect of hyperthermic temperatures is increased when the cells are simultaneously exposed to ultrasound even at intensities that do not have any effects at normal temperatures. Although these observations have not yet been reproduced in vivo (Hahn 1982), they indicate that the nonthermal interactions between ultrasonic fields and tissues may have some potential when combined with hyperthermia in cancer therapy. The importance of the mechanical interactions and stable cavitation effects may even be enhanced, if desired, by using low duty cycle and high intensity beams (total power remains unchanged). However, transient cavitation should be avoided, since at present the location and magnitude of the induced tissue damage are highly unpredictable.

2.5 Generation and Characterization of Ultrasonic Fields

2.5.1 Piezoelectric Materials

Certain materials which lack a center of symmetry in their lattice structure (i.e. they are anisotropic) have the property such that the application of pressure causes an electric voltage to appear across the crystal. The voltage is proportional to the applied pressure within the elastic limits of the material. This phenomenon is called the piezoelectric effect. Similarly, the application of an electric voltage across the crystal causes a mechanical deformation (inverse of the piezoelectric effect). In order to understand this phenomenon the crystallographic properties of these materials should be considered. In Fig. 2.28 a simple diagram of an anisotropic crystal is presented. In the crystal the positively and negatively charged particles are distributed in such a way that the total charge of the crystal is zero. When an electric voltage is applied across the crystal the positively charged particles tend to move toward the negative voltage and the negative particles toward the positive side. The crystal deforms

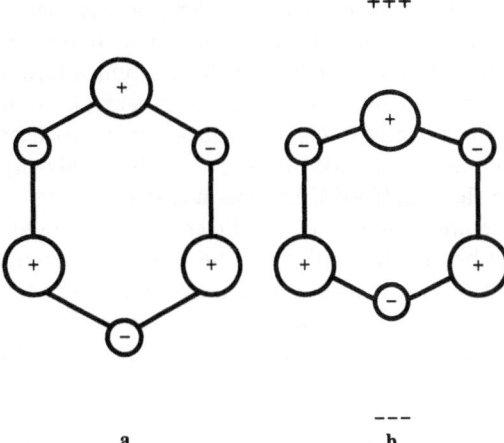

Fig. 2.28a, b. The principle of the inverse piezoelectric effect used to generate ultrasound: **a** the crystal lacking central symmetry; **b** an electric field applied across the crystal

until the elastic forces between the particles counterbalance the electric forces. Reversing the voltage induces the opposite effect. Thus, by applying a changing voltage across a piezoelectric crystal, electric energy can be converted to mechanical thickness changes of the crystal.

When the temperature of a piezoelectric material is increased and a certain − material-dependent − temperature is reached, the crystallographic structure will change to a higher order of symmetry. When this phase transition leads to a state where the material has central symmetry the piezoelectricity ceases. This temperature is known as the Curie temperature or the Curie point. For example the Curie point for quartz (SiO_2) is 573 °C and for barium titanate 120 °C.

Piezoelectricity was discovered by Jacques and Pierre Curie in natural quartz crystals in 1880. Since then, in addition to some natural crystals, a number of artificially grown crystals, e.g., ammonium dihydrogen phosphate and Rochelle salt, have been found to have piezoelectric properties. After 1956 a group of artificial piezoelectric materials known as polarized polycrystalline ferroelectrics became commercially available. After sintering, these materials contain small volumes in which the polarization is uniformly oriented. However, as a whole the material itself does not have polarization due to the random orientation of those ferroelectric domains. Polarization of the ferroelectric material is usually carried out by increasing the temperature to just above the Curie point and then allowing it to cool slowly in the presence of a strong $\overline{DC}$ electric field (typically 20 kV/cm) applied in the direction in which the piezoelectricity is desired. This polarization process aligns most of the individual charge domains.

Although the first commercial ferroelectric material was barium titanate, lead zirconate titanate (or PZT) has been the most widely used. The chemical formula of the compound is $PbTi_{1-x}Zr_xO_3$, where x is around 0.5. In the early 1970s a new group of piezoelectric materials, thin plastic membranes such as polyvinylidene fluoride (PVF_2), became available.

There are a number of coefficients that are used to describe the properties of piezoelectric materials. What follows is only a brief introduction, but for a more detailed mathematical description see for example Mason (1950), Wells (1969, 1977), and Ristic (1983).

The *piezoelectric coefficient d* describes the conversion of electric energy to strain and it is often called the transmission constant. The *piezoelectric coefficient g* (called the receiving constant) defines the electric field produced under open circuit conditions by an applied stress.

The value of the *dielectric constant* of a transducer depends on its mechanical freedom. Thus, there are usually two values that are given: ε^T when the transducer is free to move and ε^S when the transducer cannot move (i.e., when it is clamped). The dielectric constant relates the transmission and receiving constants to each other:

$$d = g\,\varepsilon^T \qquad (2.74)$$

The electromechanical (piezoelectric) *coupling coefficient k* indicates that an external radiofrequency (RF) electric field can induce acoustic vibrations in the material. Higher values of k indicate better excitation efficiency.

The *efficiency* of the transducer is defined by how much of the electric energy is converted into mechanical energy. In practice this depends on the transducer material and its mounting, and can range from 50% to 90% for ceramics and up to 99% for quartz.

In Table 2.3 the properties of some piezoelectric materials are summarized. For the induction of hyperthermia a transducer capable of producing high continuous wave output is needed. Thus, a large value of d is desired. The PZT materials appear to be superior to quartz or PVF_2 in this respect. PZT 4 and PZT 5 appear to be almost equal; however, PZT 4 has a much greater Q-factor and is thus more suitable for the generation of ultrasonic fields. Also quartz has been used for hyperthermia purposes (Lele 1983). The desirable features of quartz are that it ages little and that its properties are not temperature sensitive (it is excellent for use with lenses).

2.5.2 Ultrasonic Transducers

This section provides a short description of the operation of ultrasonic transducers for hyperthermia purposes. For further information see, for example, Wells (1969, 1977), Mason (1950), Hueter and Bolt (1955), and Ristic (1983).

2.5.2.1 Resonance Frequency

For ultrasound hyperthermia, transducers capable of producing high power, single frequency, continuous waves for extensive periods are needed. In addition to the transducer material, the mechanical structure of the transducer is very important for meeting these requirements. Consider a disk transducer (acoustic impedance Z_T) between the loading medium (water, Z_L) and the backing material (Z_B) (Fig. 2.29). When a sinusoidal continuous electric voltage is applied across the disk a stress wave enters the load and also the backing material. In addition part of the energy

Table 2.3. Electromechanical properties of some transducer materials (Wells 1969)

	Quartz, x-cut	PZT-4	PZT-5
Transmitting constant d (m V^{-1})	2.31×10^{-12}	289×10^{-12}	374×10^{-12}
Receiving constant g (V m N^{-1})	5.78×10^{-3}	26.1×10^{-3}	24.8×10^{-3}
S^E ($m^2 N^{-1}$)	12.8×10^{-12}	15.5×10^{-12}	18.8×10^{-12}
ε^T (F m^{-1})	4.0×10^{-11}	11.5×10^{-9}	15.0×10^{-9}
Wave velocity (m s^{-1})	5740	4000	3780
Density (kg m^{-3})	2.65	7.5	7.75
Acoustic impedance	1.52×10^6	3.00×10^6	2.93×10^6
Mechanical Q	>25000	500	75
Curie temperature (°C)	573	328	365
Coupling coefficient k	0.095	0.70	0.705
Wavelength at 1 MHz (mm)	5.74	4.00	3.78

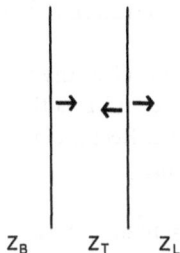

Fig. 2.29. Wave propagation in an air-backed transducer ($Z_B = 0$)

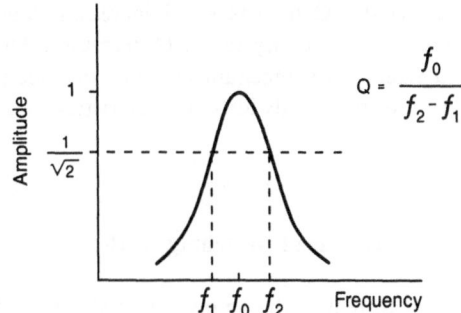

Fig. 2.30. The displacement amplitude of a transducer surface induced by a constant driving voltage as a function of the frequency around the mechanical resonance frequency (f_0)

is reflected into the transducer from both of the surfaces (unless the acoustic impedances in the medium and the transducer are equal). When the stress wave generated at the back of the disk propagates through the disk and reaches the front interface, the generating voltage has already changed and a new stress is induced. The resulting stress wave is the sum of the two waves and thus can be either smaller or larger than the original stress, depending on the phase of the waves. The phase difference depends on the thickness of the disk with respect to the frequency of the applied voltage, with the maximum stress wave obtained when the reflected wave and the generated wave are in the same phase. This situation occurs at the thickness of the plate $d = \lambda/2$ or its odd multiple. The frequency which corresponds to the half wavelength thickness is called the fundamental resonant frequency of the transducer, and it gives the maximum displacement amplitude at the transducer faces. At the frequency of 1 MHz the half wavelength is 2.9 mm, 2.0 mm, and 1.9 mm in quartz, PZT4, and PZT5, respectively. If the transducer is driven at a frequency which is three times its fundamental frequency it is operating at its third harmonic, and so on.

2.5.2.2 Backing of the Transducer

In order to maximize the energy output to the load the backing material should be selected so that $Z_T \gg Z_B$. This would cause most of the energy directed to the back of the transducer to be reflected to the front surface and the generated ultrasonic field would exit only to the load. In practice, air backing gives almost complete energy reflection from the back of the transducer.

2.5.2.3 Q-Factor

The frequency characteristic of the radiated power is determined by the mechanical quality or Q-factor of

the transducer. For the n-th harmonic mode it can be written.

$$Q_n = n\,\pi/2\,(\varrho_T\,v_T)/(\varrho_L\,v_L) \tag{2.75}$$

for an air-backed transducer. Here $\varrho_T\,v_T$ refers to the transducer and $\varrho_L\,v_L$ to the load. It can be seen from Eq. 2.75 that Q depends on the order of harmonic but not on the transducer size. In Fig. 2.30 the effect of mechanical Q is shown relative to the displacement amplitude of the transducer. The Q-factor can be determined from such a graph:

$$Q = f_0/(f_2 - f_1) \tag{2.76}$$

where f_0 is the frequency of maximum amplitude, f_1 the frequency below resonance for a reduction in amplitude of $(2)^{-1/2}$, and f_2 the frequency above resonance for a similar reduction in amplitude. The measured Q-values are lower than the calculated ones due to the mounting losses of the transducer. This can reduce the values by as much as a factor of 2. The total acoustic power emitted by the transducer is proportional to Q:

$$P = A\,Q^2 V_{rms}{}^2 \tag{2.77}$$

where A is the transducer area and V_{rms} is the driving voltage.

There is also an electric quality factor Q' which is the energy stored in the capacitance of the transducer divided by the power dissipated in R_p per cycle, where R_p is the effective parallel damping resistance of the transducer. If the crystal is loaded with liquid then

$$Q' \simeq n\pi/4k^2\,(\varrho_L\,v_L)/(\varrho_T\,v_T) \tag{2.78}$$

where k is the electromechanical coupling coefficient of the transducer material. From Eqs. 2.75 and 2.78

we can see that Q' increases with increasing load while the mechanical quality factor Q decreases. The combined effect of the mechanical Q and the electrical Q' determines the bandwidth characteristics of a transducer.

2.5.2.4 Mechanical Matching to the Load

In order to reduce the acoustic impedance difference between the transducer and the load, a matching layer can be constructed on the face of the transducer. When the impedance of the matching layer is $>Z_L$ but $<Z_T$, the layer is acting as a mechanical transformer which reflects an increased load to the transducer. The ideal thickness of the layer is $\lambda/4$ because at this thickness the wave that reflects back from the front surface of the matching layer to the transducer face and back again to the front of the matching layer is in the same phase as the wave passing through the front surface at that moment. Thus, the multiple reflections are adding to the wave at this thickness, whereas at any other thickness (except odd multiples of the quarter wave) the resulting wave from the interference would be smaller. The ideal impedance in the matching layer is $Z_M = (Z_T + Z_L)^{1/2}$, which allows all of the energy to be transmitted to the load.

2.5.2.5 Electric Matching

Maximum electric efficiency of the transducer can be obtained when the transducer is matched to the electric impedance of the driving amplifier and the electric and mechanical resonances of the transducer are tuned together. The input impedance of a crystal is given by

$$X = 1/(2\pi f C_t) \tag{2.79}$$

where C_t is the capacitance of the crystal and is

$$C_t = \varepsilon_0 \varepsilon A/d_t \tag{2.80}$$

where A is the area of the crystal, d_t the thickness of the transducer, ε_0 the dielectric constant of the free space, and ε the dielectric constant of the transducer material.

The impedance matching is often done using a transformer where the reflected impedance is proportional to the turns ratio squared. The resonant frequency f_0 of an electric system is

$$f_0 = 1/(2\pi(LC)^{1/2}) \tag{2.81}$$

where C is the total capacitance of the crystal and cable and L is the inductance of the transformer and cable. Because most of the inductance is in the transformer,

$$L = n^2 \mu_x \mu_0 A_t/l \tag{2.82}$$

where n is the number of turns, μ_x and μ_0 are the permittivities of the core and free space, respectively, A_t is the cross-sectional area of the transformer core, and l is the length of the core. By suitable choice of the transformer parameters the electric resonance can be tuned to the frequency of mechanical resonance of the transducer (see Walker and Lumb 1964). There are also several other circuits that can be used to match the electric impedance of the transducer to the electric impedance of the driving amplifier.

When the impedance of the transducer is low it is important to mount the matching and tuning network as close to the transducer plate as possible. This is because the impedances of the transducer and the cable form a potential divider, and if the cable impedance is significant compared with the transducer impedance there will be power losses in the cable. For transducers with large impedances at low frequencies the matching network can be further away from the transducer. For hyperthermia applications the electric tuning and matching of the transducer are important and large power gains can be obtained with this simple circuit.

2.5.2.6 The Structure of an Ultrasound Hyperthermia Transducer

Figure 2.31 shows the general structure of an ultrasound transducer suitable for hyperthermia treat-

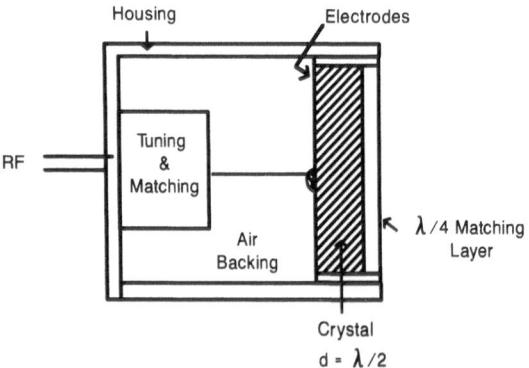

Fig. 2.31. The structure of an ultrasound transducer that can be used for induction of hyperthermia

ments. A plate of suitable piezoelectric material (e.g., PZT 4) is obtained with the thickness of half of the desired wavelength. The electric connections on both sides of the transducer are made by means of a thin layer of metal which is evaporated, sputtered, or fired on the transducer surface. The most commonly used electrode materials are nickel, silver, and gold. The wires are usually soldered on the electrodes. The transducer plate is mounted on the holder in such a way that it has maximum freedom to move. The generated sound field can be severely distorted by clamping the edges down to the holder. On the front surface there can be a one-quarter wavelength matching layer. This is optional and adequate power outputs can be obtained even without it. At the back there is an air space to provide a low impedance backing. This space can also house the matching circuit. The transducer is then connected via a standard cable to the power amplifier.

2.5.3 Calibration of Ultrasonic Fields

Before using ultrasonic sources to induce elevated temperatures in patients, the fields should be calibrated. Usually the hyperthermia systems are capable of measuring the forward and reflected electric power to and from the transducer. Thus, the operator has an idea of the total power losses in the system. In order to be able to obtain the amount of energy delivered into the tissue the total acoustic power output as a function of the net electric power should be known. In addition, the shape of the generated ultrasonic field helps to estimate the absorbed energy distribution. This is especially important with the focused fields, where most of the energy is concentrated into a fairly small focal volume. The maximum spatial acoustic intensity in the field as a function of the electric power should be known as well. This information gives the energy density or the temperature elevation capability of the field in every location during the treatment. The intensity is also important because nondesired effects of sonication – like cavitation – appear at a certain intensity level. In the following, some of the techniques that can be used to obtain the total power output, field distribution, and peak intensity will be discussed. For further information and references see Wells (1977) and Stewart (1982).

2.5.3.1 Hydrophones

Hydrophones are piezoelectric transducers that can convert local pressure variations into an electric sig-

nal. Thus, they can be used to measure the local acoustic pressure in the field, which then can be converted into intensity. Hydrophones can be calibrated to give absolute pressure values by measuring the electric output of the hydrophone in a known ultrasonic field. If the sensitive element size is smaller than one-tenth of the wavelength, the directivity is sufficiently uniform for most practical purposes. Larger detectors (which are normally used in the MHz range) are directional and thus can cause severe distortions in the measured field patterns, especially in strongly focused fields. In addition, the hydrophone itself interacts with the sound field, causing standing waves and measurement errors. These errors are most prevalent with ceramic-type detectors. Hydrophones made of thin, piezoelectric polymere membrane (PVF_2) have an acoustic impedance much closer to water than the ceramics and are almost transparent to the ultrasonic beams. The detector size can be made as small as 0.5 mm in diameter with a fairly flat frequency response. An alternative method used to reduce the disturbance of the ultrasound field by the hydrophone is to scan a small spherical scatterer in the field and measure the amount of scattered ultrasound with a large focused hydrophone located outside the field in such a manner that the sphere is always in the focus of the hydrophone (Edwards and Jarzynski 1980).

2.5.3.2 Radiation Force Measurements

Radiation force measurements are probably the techniques most commonly used to obtain the total power output from the transducer and the local absolute intensity in the field. The methods are based on the momentum change that occurs when an ultrasonic beam is reflected or absorbed.

Total Power Measurements. Let us consider first the total acoustic power measurements. The force produced on the reflecting or absorbing target is proportional to the total ultrasonic power (P) intercepted by the target surface. The radiation force (F) is

$$F = DP/v \qquad (2.83)$$

where v is the propagation velocity of the wave in water and D is a constant that depends on the type of interaction between the target and beam. For a perfect reflector and absorber at normal incidence D = 2 and 1, respectively. Some values for different types of target are given in Table 2.4. The measurement using reflective targets is fairly sensitive to the accuracy of the angle between the beam and the reflector surface,

Table 2.4. D-values for the radiation force measurements in different situations (Stewart 1982)

Target	θ	ϕ	δ	D
Absorbing	0 – 90	0		1
	0 – 90	ϕ		$\cos \phi$
Reflecting	0	0	0	2
	45	0	0	1
	θ	0	δ	$2 \cos \theta \cos \delta$

θ, angle of incidence; ϕ, angle between the beam and the direction of the force measurement; δ, angle between the normal of the surface of the target and the direction of the force measurement

and all measurements depend on the angle between the beam axis and the direction in which the force is measured. The situation is more complicated with focused fields. If R is the radius of curvature of the ultrasonic transducer, L is the focal length, and the angle of incidence is 0, then the force measured along the beam axis is

$$F' = (2P/v)(2L/R^2)[(R^2+L^2)^{1/2}-L] \qquad (2.84)$$

For an absorbing target it can be written

$$F' = (P/v)(2L/R^2)[(R^2+L^2)^{1/2}-L] \qquad (2.85)$$

(Stewart 1982).

When absorbing targets are used care should be taken that there is very little reflection back from the target to the transducer. This would cause standing waves and, hence, error in the force measurement.

There are several ways in which the radiation force can be measured in practice. One of the most accurate and convenient methods is to hang the target down from a microbalance and direct the ultrasonic beam to the target from straight above or below (Fig. 2.32). The radiation force can now be obtained from the weight change (m) observed by the balance

$$F = mg \qquad (2.86)$$

where g is the gravitational constant ($= 9.81 \text{ ms}^{-2}$). With high power levels the absorbing target tends to expand and thus its weight changes during sonication. Therefore, the measurements should be performed fairly rapidly after the power is turned on, or pulsed (instead of continuous wave) ultrasound used to reduce the absorbed power (Fig. 2.33). Highly accurate readings can be obtained by linking the microbalance into a computer which can perform the measurements fast and correct the errors caused by

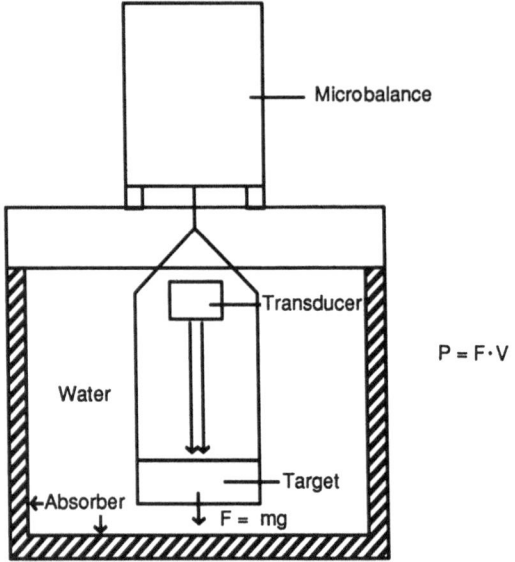

Fig. 2.32. The total acoustic power measurement using radiation force detection

the heating of the target. The streaming of water induced by the sound field can also cause some error in the force measurement.

Radiation force measurements are easy to perform; the readings are accurate (within 5% – 10%) and the technique is absolute, i.e., there is no need for calibration. In addition, it is fairly easy to construct a good measurement system without great expense.

Intensity Measurements. The radiation force exerted on a small sphere suspended in an ultrasonic field is proportional to the average intensity over the cross-section of the sphere. When the acoustic properties of the sphere are known, the absolute intensity can be calculated from the force. Most often the spheres are made of stainless steel. In a typical measurement system the sphere (small compared with the field variations) is suspended in the desired location by thin

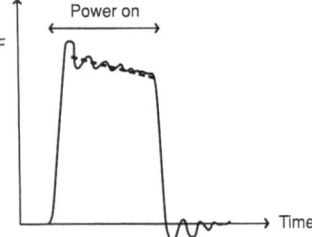

Fig. 2.33. The measured radiation force on an absorbing target during high power sonication. Note the decrease in the measured force due to the heating of the target

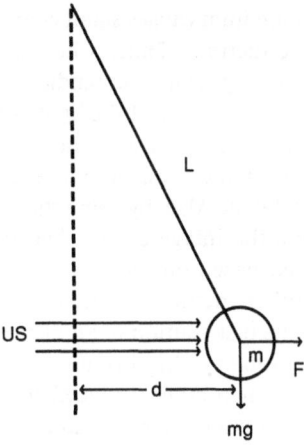

Fig. 2.34. The acoustic intensity measurement utilizing the radiation force on a small steel sphere

nylon fibers (Fig. 2.34). The radiation force (F) deflects the sphere a distance (d). From the measurement of the deflection the radiation force can be calculated:

$$F = mgd(L^2 - d^2)^{-1/2} \qquad (2.87)$$

where L is the length of suspension, m is the total mass of the sphere and the thread in water, and g is the gravitational constant. The absolute average intensity over the sphere can be calculated from the radiation force

$$I = Fv(a^2Y)^{-1} \qquad (2.88)$$

where a is the radius of the sphere, v is the speed of sound in water, and Y is the radiation force function. Y depends on the ratio of the wavelength in water to the radius of the sphere and also on the temperature and sphere material. For a stainless steel sphere (440 C) accurate Y values can be obtained from the literature (Stockdale and Hill 1976; Dunn et al. 1977).

2.5.3.3 Thermal Methods

Calorimetric techniques can also be used to obtain either the total power output of a transducer or the absolute local intensity (or pressure amplitude) in the field. This technique is not affected by the beam shape (i.e., it is omnidirectional) or the pulse duration.

Total Power Output. The total ultrasonic power output from a transducer can be measured by directing the beam into an adiabatic chamber filled with absorbing fluid. By measuring the temperature elevation

of the fluid during the exposure the total power output can be calculated. When done properly, the calorimetric approach is the most accurate method of measuring the acoustic power output. The main disadvantage of the earlier systems was that they took long times to return to equilibrium and thus, multiple exposure measurements were very slow. To avoid this and to obtain a fast measure of the acoustic power, dual-chamber flow-through calorimeters have been developed. In this approach fluid is constantly circulated through two identical chambers and a heat exchanger. The ultrasound beam is directed into one chamber, thereby increasing the temperature of the absorbing fluid which is measured in the outlet. In the other chamber a small electric heater is immersed and the power through it is increased during the sonication until the fluid temperatures coming out from both of the chambers are equal. The ultrasonic power is equal to the applied electric power. Because the fluid temperature is returned to the base line in the heat exchanger, the system returns to equilibrium very fast. This type of calorimeter can be made very accurate.

Local Intensity Measurements. Local ultrasonic intensity can be obtained by measuring the rate of temperature rise in an absorbing medium during sonication. Fry and Fry (1954a, b) embedded a thermocouple probe in a thin castor oil cell (Fig. 2.35) and directed an ultrasonic beam to the thermocouple via a thin plastic membrane window. During a 1-s sound pulse the temperature first increased fast and then after 100 ms or so started to elevate linearly (Fig. 2.35). The initial temperature rise is caused by the viscous forces acting between the thermocouple wire and the oil and the linear rise by the energy absorption in the oil. By determining the slope of the temperature rise around 0.5 s from the beginning of the pulse, the absolute intensity can be calculated:

$$I = (\varrho c/2\alpha)dT/dt \qquad (2.89)$$

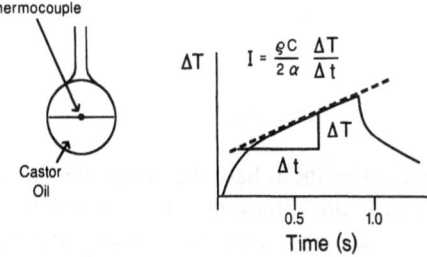

Fig. 2.35. The castor oil cell for intensity measurements and the temperature increase during the sound pulse measured with a thermocouple (Fry and Fry 1954a, b)

Table 2.5. Acoustic properties of castor oil (Fry and Dunn 1962) at the frequency of 1 MHz

Temperature (°C)	Density (kg m^{-3})	Speed (m s^{-1})	Absorption coefficient[a] (Npm^{-1})
0	972	1580	26
10	960	1536	16
20	952	1494	9.6
30	946	1452	5.7
40	941	1411	3.7

[a] The absorption proportional to (frequency)$^{5/3}$ at 30 °C

where ϱ is the density, c is specific heat, and α is the amplitude absorption coefficient of the oil. In Table 2.5 the values for castor oil are given. At longer sonication times or with small diameters of the beam the temperature elevation starts to bend from the linear increase due to the effect of thermal conduction. The temperature increase also has an effect on the absorption coefficient of the oil. This causes an error in the measurement when large intensities are used. Similarly the fluid convection becomes important at high intensities. These problems can be partly avoided using a polyethylene absorber and mathematical correction techniques to obtain the intensity (Yoshioka and Oka 1965).

The same principle can be used to obtain intensity distributions. However, instead of using a large volume of an absorber, a thermistor or a thermocouple can be coated with a small sphere of an absorbing material (Martin and Law 1983). Now the rate of temperature rise during the very beginning of the sound pulse provides a measure of the intensity. This parameter is linearly proportional to the intensity and the measurements can be repeated fast enough to map the whole ultrasonic field with adequate accuracy. The main advantages of this technique are its simplicity, i.e., a good measurement system is easy to build, and the fact that the probe is omnidirectional, thus providing the true energy distribution in the medium. This technique gives only relative field distributions unless the probe is calibrated against another measurement technique.

2.5.3.4 Optical Methods

Optical methods have the advantage that they do not perturb the ultrasonic fields during measurements. There are several phenomena that can be utilized to obtain the ultrasound field characteristics via optical means (see Harran 1977; Cook 1977). One can utilize the fact that acoustic pressure in an optically transparent medium causes small changes in the optical index of refraction. Thus, when a monochromatic light beam is passing perpendicularly through an ultrasound field, part of the light is diffracted by the ultrasound. From the amount of diffracted light the integral of the acoustic pressure over the beam can be calculated. Also, by using the schlieren (shadow) technique the image of the field distribution can be obtained based on the diffraction.

Another method of measuring the acoustic pressure and its distribution in an ultrasonic field is optical interferometry. In this method a thin (8-μm), gold-plated membrane is placed across the ultrasonic field. Then a laser beam is scanned over the plate to obtain the particle displacements caused by the ultrasonic field. The main advantages of this technique are high sensitivity, large dynamic range, wide angular response, good resolution, and broad frequency response.

Despite the many advantages of optical methods, the measurement systems are not yet widely used. This is perhaps due to the lack of commercial devices and to their fairly high costs compared with the other methods, which often produce adequate results.

2.5.3.5 In Vivo Measurements of Ultrasonic Fields

There are several factors that affect the shape and magnitude of an ultrasonic beam when it propagates in tissue. First the beam is attenuated by absorption and scattering, second it is reflected back at interfaces, and third the heterogeneity of different tissues may alter the beam shape and its propagation direction. Thus, it is important to measure ultrasonic fields in vivo. So far there is little information about intensity measurements in vivo. Martin et al. (1984) used a coated thermocouple probe to obtain intensity distributions in several tissues. One can also use uncoated thermocouples but then the absorption differences in the tissue can have an effect. Another method is to use a small hydrophone probe at the tip of a needle. At present the intensity measurement techniques are fairly complicated to apply in vivo and thus they have not been used in clinical treatments.

2.5.3.6 Characterization of Ultrasonic Fields for Hyperthermia Treatments

Before using ultrasonic transducers for inducing hyperthermia they should be calibrated. The calibration should be verified periodically, though ultrasonic

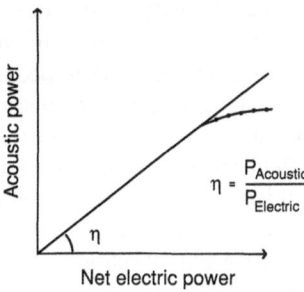

Fig. 2.36. The acoustic power as a function of net electric power (= forward − reflected). The *dotted line* indicates the nonlinear response, which may indicate that the acoustic power output is not stable

that the transducer can be operated at. (This operational power range usually gives a linear relationship between the electric and acoustic power and the nonlinearity is often an indication of a power usage that would lead to an unstable power output and possible damage in the transducer or the driving circuitry.

Second, the spatial peak intensity in the field as a function of applied net electric power should be measured similarly to the total acoustic power. Again, the calibration should be done throughout the whole operating range and the calculated intensities (with attenuation correction) displayed during the treatment. Third, the ultrasonic field distribution should be measured in a plane including the central axis of the ultrasonic field. Also the field distribution across the beam in the focal plane would be useful (Fig. 2.37). If several overlapping beams are to be used, the field measurements together with the peak intensity measurements should be repeated with the whole array of the transducers. These measurements should be done first in water and then repeated later in a phantom having an attenuation coefficient equal to the one in tissue. The phantom measurements will be discussed in the following sections.

transducers are fairly stable. Three pieces of information are needed for proper characterization of an ultrasound source. First, the total power output should be measured as a function of the net electric power (Fig. 2.36) at several points covering the whole range of powers to be used. This gives the efficiency of the transducer and allows the calibration factor to be input into the computer programs reading the electric powers, and thus the total acoustic power can be displayed. It also gives the maximum electric power

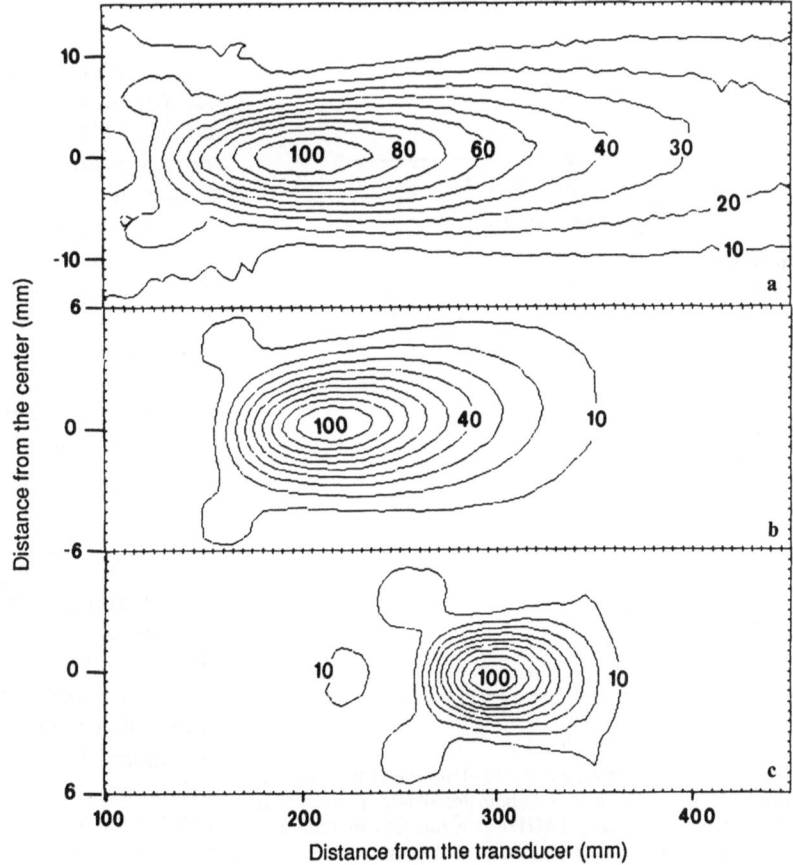

Fig. 2.37a−c. The acoustic pressure amplitude² distributions measured in the axial plane of three different focused transducers in water

2.5.4 The Use of Ultrasonic Phantoms

There are several phantoms that can be used to mimic ultrasonic and/or thermal properties of tissues. Phantoms can be used: (a) to obtain ultrasonic field distribution experimentally, (b) to optimize ultrasonic transducers, (c) to test theoretical models and (d) control algorithms, and (e) to test equipments. The phantoms vary from a simple water phantom to a complex perfused phantom depending on the purpose for which the phantom is to be used.

2.5.4.1 Liquid Phantoms

Liquid phantoms are mainly used for quality assurance purposes to obtain the ultrasonic field distributions, the total power output from the transducer, or

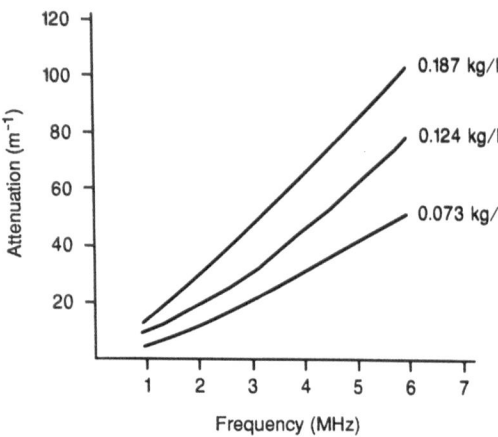

Fig. 2.38. The ultrasonic attenuation as a function of frequency in a gelatin-graphite phantom for different concentrations of graphite. (Reprinted with permission; Madsen et al. 1978; © 1978 Pergamon Press plc.)

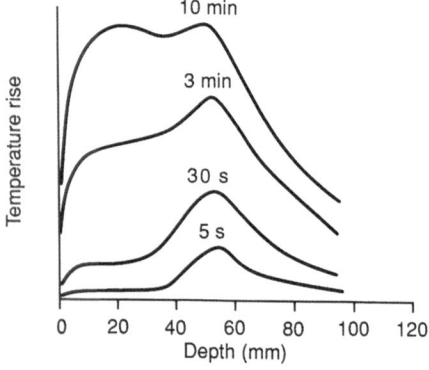

Fig. 2.39. The temperature increase along the central axis at a stationary focused transducer (diameter 38 mm, radius of curvature 70 mm, frequency 1 MHz) as a function of time in an agar-graphite phantom (Hynynen et al. 1983a)

the peak intensity in the field. The most common phantom is a water phantom which has the speed of sound and an acoustic impedance close to those of tissues. This allows the measurement of the total acoustic power output from the transducer and also the field measurements without attenuation. The attenuation can be taken into account by adding it mathematically or using absorbing liquid instead of water. By a suitable choice of the fluid and its temperature, the tissue attenuation can be simulated. Liquid phantoms are very useful since one can map the field at any points in the phantom.

2.5.4.2 Solid Phantoms

There are several solid phantoms that have been developed to test diagnostic ultrasonic equipment. These phantoms can simulate the acoustic properties of various organs and even anatomically accurate phantoms have been built. The most common technique to obtain the right attenuation and scattering is to mix small particles (like graphite or talc powder) in a gel (gelatin or agar) (Fig. 2.38). The speed of sound can be varied by mixing a small amount of alcohol with the water. In some other phantoms the absorptions has been increased by mixing oil or albumin with the gel (Lele and Parker 1982; Madsen et al. 1982). With these phantoms even the thermal properties (thermal conduction and diffusivity) can be mimicked. The difficulty with solid phantoms is that the temperature sensors have to be implanted in discrete locations in the phantom and often permanently implanted thermocouples have been used to obtain the temperature distributions. Solid phantoms can provide the heating field distribution, similarly to the liquid phantoms, and in addition the beam disturbances caused by various tissue layers if anatomically accurate phantoms are used. However, the temperature distributions measured in these phantoms after long sonication times do not have great significance since the perfusion modifies the distributions greatly in tissue in vivo (Fig. 2.39).

Finally, there have been some attempts to use postmortem tissue samples as phantoms for ultrasound hyperthermia. Due to the microbubble formation after the death of the animal, the ultrasonic attenuation of the tissue increases severalfold within 2 days, making it difficult to simulate the in vivo properties (Fig. 2.40) (Bamber and Nassiri 1985). This effect can be reduced by degassing the samples under vacuum before use. Thus, great care should be exercised when tissue samples are used as phantoms for ultrasound hyperthermia.

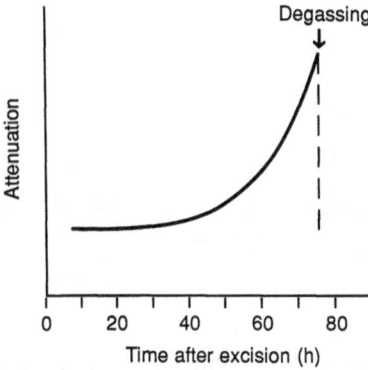

Fig. 2.40. The ultrasonic attenuation measured in porcine liver as a function of the time after excision. (Reprinted with permission; Bamber and Nassiri 1985, © 1985 Pergamon Press plc.)

2.5.4.3 Perfused Phantoms

Perhaps the most important heat transport mechanism contributing to the induced temperature distributions during hyperthermia treatments is the blood flow in large vessels and the perfusion through the capillary network. There have been a couple of attempts to simulate the perfusion in ultrasonic hyperthermia phantoms. Edmunds et al. (1985) reported preliminary results from a phantom where the perfusion was accomplished by implanting a large number of small polyethylene tubes in a gel-based phantom. Similarly, Chin et al. (1986) proposed using small (between 0.6 and 3.5 mm) spheres made of phantom material and packed closely to simulate tissue and then simulating the perfusion by pumping 10% n-propanol between the spheres. However, at present there is no experimental evidence of how well this approach works.

An alternative approach is to use alcohol-fixed organs and pump water through them (Holmes et al. 1984). Providing that the acoustic properties of the organ are known and the vascular network remains intact, this type of a model may offer the best in vitro experimental medium to test ultrasound hyperthermia systems.

2.6 Ultrasound Systems for Induction of Hyperthermia

The following section is a summary of the existing hyperthermia devices that utilize ultrasound as the method of heating. A detailed technical description of the various pieces of equipment will be avoided and only the system characteristics that are important for the understanding of their function will be given. The generation of the RF signals to be converted into mechanical motion is in principle similar in all systems; therefore, a typical system diagram is presented in Fig. 2.41. The RF signal is generated by a signal generator or an oscillator and is amplified by an RF amplifier. Commercial amplifiers and frequency generators have been used in many of the laboratory systems. The forward and reflected electric power is measured after amplification in order to obtain the total acoustic power output (the system must be calibrated to give acoustic power as a function of net electric power). Before the signal enters the crystal, it passes through a matching and tuning network that couples the electric impedance of the transducer to the output impedance of the power amplifier. This matching network can be in the transducer head itself or at some distance away in the line if the transducer impedance is high compared with the cable. The power output can be controlled either by using a fixed gain amplifier and controlling the level of the input signal from the frequency generator or by controlling the gain of the amplifier. Also, a fixed gain system can be used with a controllable duty cycle. In many of the current laboratory systems the power output is manually controlled but in the future systems computer-controlled power will most likely be an option.

In a hyperthermia device a temperature measurement unit is needed to monitor and record the temperatures. In most of the ultrasonic systems, thermocouple probes have been used owing to their small size. The signal from each thermocouple can be read in sequence by using a multiplexing system and a digital

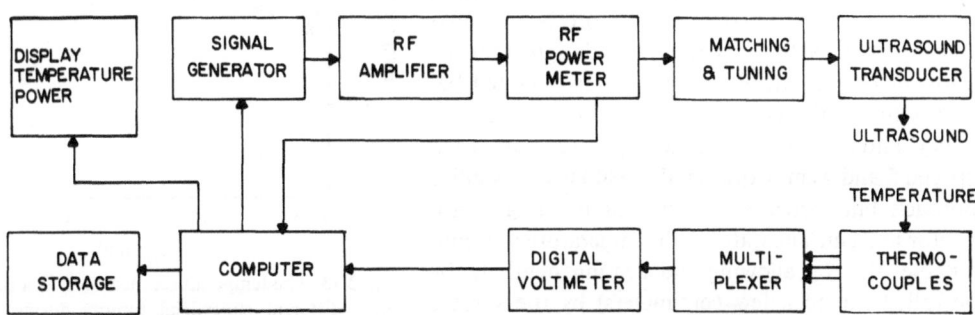

Fig. 2.41. An ultrasound hyperthermia system

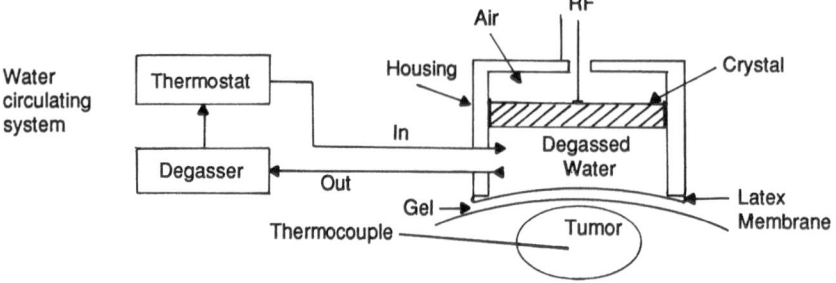

Fig. 2.42. A planar sonication head used in superficial hyperthermia treatments (Marmor et al. 1979)

voltmeter. There are also commercial systems that can be used with thermocouples. In order to read an adequate number of temperatures a computer is often used to execute the measurement and store the temperatures. The temperatures should be read when the sound is periodically turned off in order to avoid temperature measurement artifacts associated with the probes in the ultrasonic field.

A clinical ultrasound device aimed at the treatment of deep tumors also requires some way of planning the treatment. The anatomical information from CT or ultrasonic scans can be utilized to design the beam angle, the frequency, and other characteristics of the treatments. This requires software and hardware capable of transferring the anatomical information into the computer memory. To some degree this treatment planning can be done just by considering the geometrical factors (see Lele and Goddard 1987), but ultimately it would be useful if the thermal properties of the tissues could be taken into account.

rangement, the tumor was located in the near-field of the transducer, where the average power output is fairly uniform across the applicator face (thermal conduction smooths the near-field interference maxima and minima). As expected, this uniform energy input produced fairly good temperature distributions in the tumor, though the temperatures at the edge were usually lower than at the center. Similarly, the temperatures as a function of depth were fairly good (Fig. 2.43). This type of an applicator has recently been used in the induction of hyperthermia intraoperatively (Coughlin et al. 1987). The temperature distributions achieved in this clinical trial are promising.

In order to obtain equal temperatures at the periphery of the tumor and the center, more energy needs to be deposited at the edge. There have been two ways to do this. First, Munro et al. (1982) designed a transducer where the crystal is divided into two rings, each driven with separate power sources. By applying more power to the outer ring the conduction effects at the outer edge of the heated field can be compensated for and

2.6.1 Planar Transducer Systems

The first clinical ultrasound systems utilized single circular and planar transducers, which were sonicating through a temperature-controlled water column to the patient. The applicator was sealed with a thin latex membrane that could follow the surface contours and it was coupled to the skin with acoustic gel. In Fig. 2.42, a diagram of the sonication head used in the patient treatment system at Stanford is presented (Marmor et al. 1979). This device was designed for the treatment of superficial tumors; it was operated between 1 and 3 MHz (separate transducers for each frequency) and the diameter of the applicator was varied between 2 and 4 cm. Corry et al. (1982) used a similar approach but increased the size of the largest applicator to 5 cm. The water bath temperature was controllable, thereby allowing the heating depth to be controlled (up to a few centimeters) by the surface temperature and the frequency. With this kind of ar-

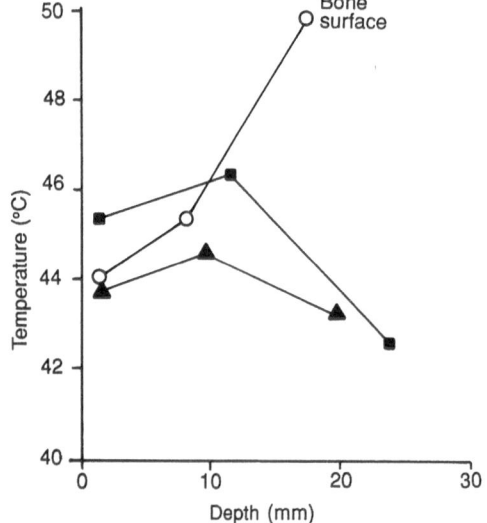

Fig. 2.43. The temperatures measured as a function of depth in three different superficial tumors during sonication with a planar ultrasonic transducer (data from Marmor et al. 1979)

good temperature distributions can be produced. The second approach was developed by Lele (1981): A lens producing a ring focus around the tumor and a secondary focus at the center of the tumor was utilized. This was found to produce good temperature distributions in clinical patient trials.

There are now several commercially available clinical ultrasound systems that utilize the single plane ultrasonic applicators. These systems appear to work fairly well when the tumors are small. In order to heat larger tumors, multielement applicators have been designed with independent power input to each transducer element. This allows a variable power output over the heated area to compensate for variations in the cooling by blood flow and thermal conduction, and to adapt to the geometry of the tumors. A 16 square element (4×4) array described by Underwood et al. (1987) has been tested clinically and is now commercially available (Fig. 2.44). The individual element size is 36 mm×36 mm and the operating frequency is 1 or 3 MHz. Since the tumor is always within the nearfield of the transducer, the beam is well collimated and propagates to the volume in front of each element. This allows good control over the power deposition pattern. This kind of approach appears to add flexibility and control over the heating pattern and produces better heating than single element applicators. Clearly, there is room for further development of this kind of approach, such as making the applicators more practical for the clinical environment.

2.6.2 Multiple, Overlapping Nonfocused Fields

In order to overcome the effect of attenuation and to deliver more energy into a deep tumor, several plane transducers with beams that are overlapping at the depth in question can be used. The first system that was clinically evaluated was constructed from six 350-kHz circular plane transducers (diameter 70 mm) which were mounted on a spherically curved surface (radius of curvature 260 mm and aperture angle $\pi/2$) in such a manner that the angles between the transducers were 60° and the convergence angle (central axis of the beam to the central line of the array) could be varied between 25° and 30°. Because the beams could be independently aimed they could be adjusted to all overlap at one location or to form a larger, more spread out focus (Fessenden et al. 1984) (Fig. 2.45). This system has been tested both in animals and in clinical trials and it was found to be

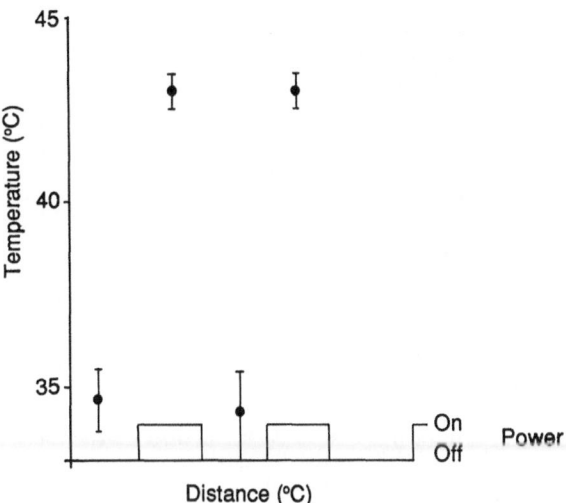

Fig. 2.44. The temperatures measured in dog's thigh in vivo during sonication with a row of four ultrasonic transducers of which two were off and two on (data from Underwood et al. 1987)

able to heat tissues effectively at depth. In 24 of the 57 clinical treatments, therapeutic temperatures were reached. However, this system is not used in the clinic presently, owing to the large number of treatments (30%) limited by pain. The reason is obvious, i.e., the low frequency used (350 kHz) penetrates deep into the tissues, propagating beyond the focal region. Thus, bone or air interfaces will create hot spots in areas beyond the target volume. At present, the multiple stationary beam systems utilize focused transducers, which offer better control over the power deposition pattern along the beam path.

2.6.3 Focused and Stationary Fields

As was discussed in Sect. 2.2.2, an energy maximum can be achieved at any practical depth in soft tissues by using a suitable focused ultrasonic field. The focal size is determined by the radius of curvature, diameter, and frequency of the transducer. However, with most transducers that are practical for hyperthermia purposes, the focal size is too small to cover the whole tumor. For this reason, multiple fields are needed to heat a larger volume and allow better control over the power deposition distribution in the target volume. One of the clinical situations in which single focused beams have been used is in the treatment of tumors of the eye. Lizzi et al. (1984) developed a spherically curved transducer (diameter 40–80 mm, radius of curvature 80 mm) that operated between 4 and 10 MHz. The larger spherical transducer had a central

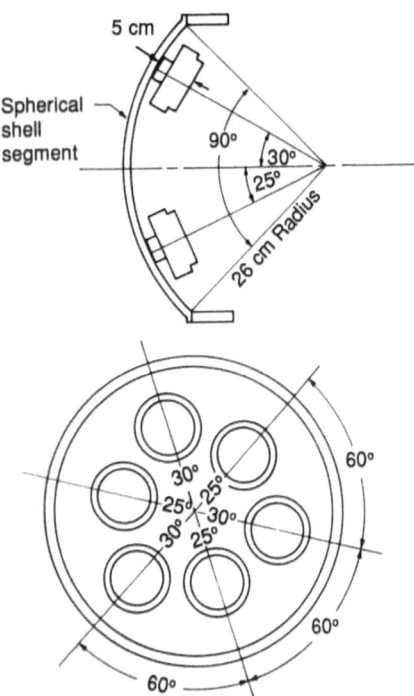

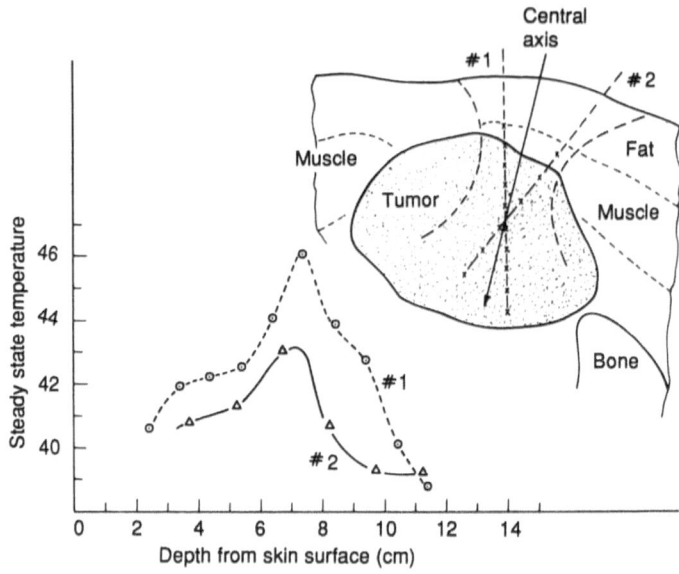

Fig. 2.45. The Stanford planar transducer array (*left*) and the temperatures measured during a clinical patient treatment (Printed with permission; Fessenden et al. 1984, © 1984 IEEE)

hole through which a diagnostic ultrasound transducer could image the target volume. This technique has already been used in the treatment of some clinical tumors (Coleman et al. 1986).

There have been several devices developed that utilize multiple focused beams. One such device was a seven focused beam system, where the central ultrasonic transducer was symmetrically surrounded by the other six and the angle of each element was variable. The operating frequency was 1.1 MHz and the radius of curvature and diameter of each of the transducers were 120 mm and 50 mm, respectively. This array was tested in pig thigh muscle in vivo and was shown to be able to produce the maximum temperature at regions deep in the tissue. Also, by offsetting the individual foci, a larger volume could be heated. This array, combined with a computerized feedback control system (Hynynen et al. 1983 b), was also tested in a clinical trial.

The same idea has been further extended by Seppi et al. (1985), in a system that utilize 30 focused beams in an array. The transducers are arranged in a symmetrical spherical pattern of four concentric rings containing 6, 6, 6, and 12 transducers each (from center to outer edge). Each ring is mechanically driven by a stepper motor which causes the transducer to pivot and thus a focal ring can be generated up to a distance of 32 cm from the transducers. The ring diameter can be varied between 0 and 10 cm. The electric power into each transducer can be independently controlled and the frequency can be varied between 0.5 and 1 MHz. The beams are focused by conical plastic lenses which produce a line focus (i.e., a long and narrow focus). The whole array is immersed in a water bath, on top of which the patient lays. The transducer gantry also includes a diagnostic ultrasonic transducer used for imaging the target volume. Although this system has not been tested in clinical treatments, it has the potential of heating some deep tumors at certain sites.

2.6.4 Focused and Scanned Fields

Another approach used to increase the size of the heated volume is to scan the transducer in such a manner that the focus travels throughout the whole tumor. This allows good control over the power deposition pattern, since the power can be controlled as a function of the focal location. Thus, the power output can be tailored for each tumor to give the desired temperature distribution, provided that temperatures are measured in an adequate number of locations. A significant advantage of this method over many of the current hyperthermia techniques is that because the scanning is usually executed under computer control, the shape and size of the treated volume can be accurately controlled.

Historically, the first scanning system (Lele 1975) was a mechanical scanner and so far this is the only type used in patient treatments. However, the scanning also can be done electrically (see Sect. 2.2.2), which probably will allow for less expensive devices after the initial development and testing. In theory a desired power deposition pattern, similar to those obtained with scanned fields, can be produced by utilizing the field conjugation method commonly used in optics (Ibbini and Cain 1988). In this technique, the phase and the amplitude of the drive signal are adjusted to the values necessary to synthesize the desired heating pattern. This is established by using the reciprocity theorem combined with the phase conjugation and amplitude matching technique.

At present there are only two mechanical scanning devices under clinical trial (Lele 1983; Hynynen et al. 1987a), and no devices are commercially available. The temperature distribution achieved during scanned focused ultrasound hyperthermia depends on several factors, as summarized in Table 2.6. In the sections to follow some of these parameters will be discussed in more detail in order to help the reader gain an understanding of their importance and to give some rough guidelines on obtaining adequate therapy for different treatment situations. This discussion is mainly aimed toward mechanically scanned systems, but can be applied to electric systems as well.

2.6.4.1 Scanning Speed

During scanned focused ultrasound hyperthermia, the focus is moved at a desired speed, over a repetitive pattern, in such a manner that the delivered energy produces a therapeutic temperature elevation in the target volume. Thus, the energy input into a tissue volume within the scanning path is periodic, consisting of a short pulse of energy followed by a longer period without any energy input. This results in a

Table 2.6. Some of the factors affecting the temperature distribution during scanned, focused ultrasound hyperthermia

1. Transducer: diameter, radius of curvature, frequency, number of transducers
2. Scanning: speed, pattern, depth
3. Tissue properties:
   a) Thermal: blood perfusion rate, blood flow in larger vessels, thermal conduction
   b) Ultrasonic: attenuation, absorption, speed, density
4. Tissue interface: bone, gas
5. Nonlinear propagation: high intensities
6. Cavitation: high intensities

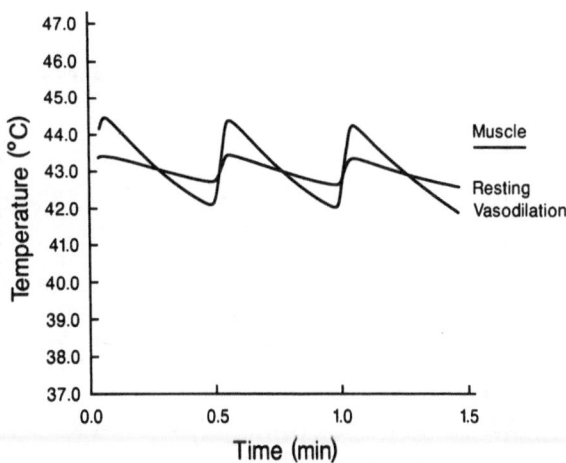

Fig. 2.46. Simulated temperature fluctuations in a resting and vasodilated muscle as a function of time during a single 20 mm diameter scan with a scan period of 30 s

sharp temperature increase when the beam passes the location and a subsequent temperature decay while other parts of the target volume are scanned. The magnitude of the temperature fluctuation clearly depends on the scanning period, i.e., the time the beam takes to complete the scan path. These fluctuations can vary from practically nonexistent when the period is very short to a situation in which the temperature decays to the baseline before the next energy input occurs. Figure 2.46 shows an example of the temperature fluctuations for a 30-s scan period in a simulated tissue with two different perfusion rates. It is obvious that the temperature fluctuations are also perfusion rate dependent, being larger in a tissue with a higher perfusion rate. In addition, the magnitude of the fluctuations depends on the focal diameter, being larger for smaller focal diameters.

The computer simulations (Moros et al. 1988) and in vivo animal experiments (Hynynen et al. 1986) have both shown that these temperature fluctuations happen only at the locations where the beam passes, and that they are important only from the point of view of the given thermal exposure (average temperature during the scan cycle is not scanning speed dependent). For scanning periods less than 10 s, the exposure given by the scanning is practically the same as the exposure given by the average, nonfluctuating temperature. This applies for most of the tumor perfusion values tested (Figs. 2.47, 2.48). In many cases the scan period may be as long as 30 s without having a significant effect on the thermal exposure.

Perhaps a more important point than that of the thermal exposure during the temperature fluctuations is that if the normal tissue temperature maximum dur-

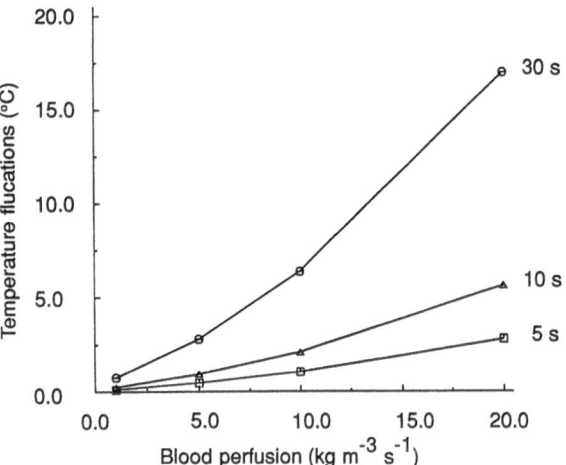

Fig. 2.47. The temperature fluctuations on a point on the beam path during a scan cycle as a function of blood perfusion and scan time (Moros et al. 1988)

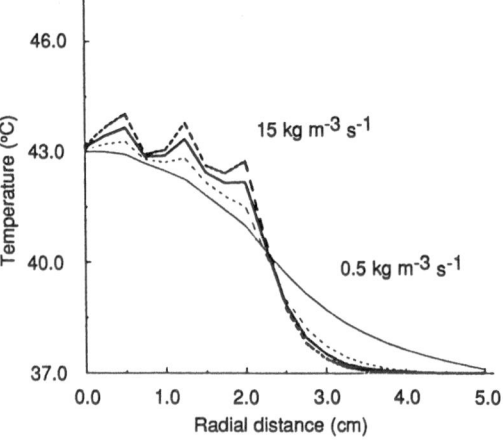

Fig. 2.49. The steady state temperature distribution along the scan radius at the focal plane with various perfusion rates (between 15 and 0.5 kg m^{-3} s^{-1}) when the spacing between the scans was the diameter of the focus (Moros et al. 1988)

2.6.4.2 Scanning Pattern

One of the first questions to be answered when planning a hyperthermia treatment using a scanned focused ultrasound system is what kind of scanning pattern should be used. The first animal experiments (see Lele 1983 for review and Hynynen et al. 1986, 1987a) have shown that the scanning pattern required to obtain a uniform temperature distribution in the whole scanned volume depends on the tissue perfusion rate. In a tissue with low perfusion a fairly large spacing between the scans can be used, and more energy is required at the edges of the heated region than at the center in order to compensate the increased thermal conduction effects at the edges. A simulation study has verified these trends (Moros et al. 1988) and has also shown that the optimum spacing between scans (concentric scans in the study) is less than or equal to the diameter of the beam focus (Fig. 2.49). This kind of scanning pattern would give a fairly uniform temperature elevation, across the scanned volume, at all of the practical perfusion rates. However, if the perfusion is small then the temperature along the edges will be lower than that at the center. In addition, any perfusion variations inside the scanned volume will create a nonuniform temperature distribution. Each of these effects can be compensated for if the temperatures in various locations in the target volume are measured during the treatment and then used to control the power output as a function of location. As a summary, when the perfusion rate distribution is unknown, optimum heating can be provided by using a multipoint feedback controller and a scanning pattern

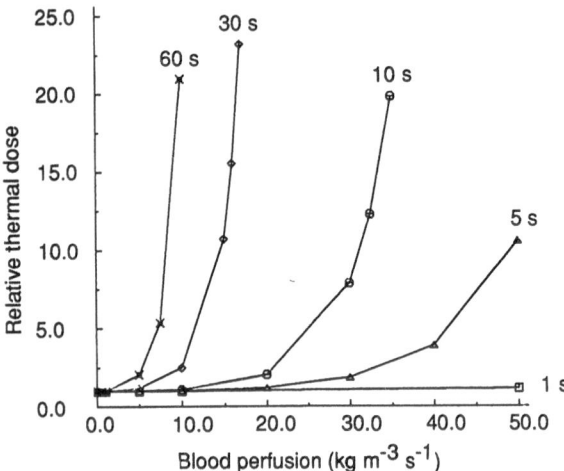

Fig. 2.48. The relative thermal dose (= thermal dose of the fluctuating temperatures/thermal dose of the average temperature) to a point on the scanning path as a function of blood perfusion (Moros et al. 1988)

ing the fluctuations is close to the pain threshold (44°–45 °C), then slow scanning can cause periodic pain during the treatment (Hynynen et al. 1987b). This may be especially important if the beam is propagating to a bone or skin surface.

There has also been speculation that the temperature fluctuations caused by slow scanning might be advantageous for giving an increased thermal exposure when they occur inside the tumor volume (Hynynen et al. 1986). Presently, there is no experimental evidence from animal or human tumors to support this.

where the scans are separated by the distance of the beam diameter. This applies at the focal plane.

The size of the scan has a significant effect on the resulting temperature distribution, particularly when deep tumors are heated. This is especially true if the transducer is scanned with a single beam oriented normal to the surface. The reason for this is that the ratio of the surface area covered by the scanned beam to the target area decreases with the increasing scan size, being finally less than the losses caused by attenuation if the transducer is not large enough. Another disadvantage of this type of scanning is that some parts of the beam are at all times passing through tissues along the central axis, both in front of and behind the focal area. This causes higher temperatures in these regions (mainly in front of the focus) as compared with those at the focal plane (Fig. 2.50). The magnitude of this effect depends on the F-number of the transducer, the frequency, the diameter of the scanning path, and the focal depth (see Moros et al. 1989 for further detail). A way to avoid this problem is to tilt and rotate the transducer (Lele 1983) in such a manner that the beam is not overlapping with the scan axis (Fig. 2.51). This problem can also be reduced by using multiple beams and switching off each beam when it passes through the central part of the scan to the opposite side (Moros et al. 1989) (Figs.

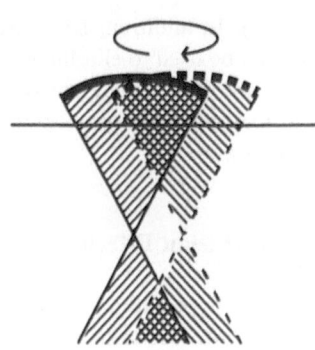

Fig. 2.50. The beam overlap when a focused beam is scanned

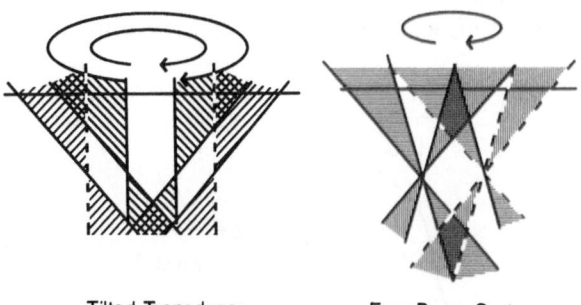

Tilted Transducer   Four Beam System

Fig. 2.51. The beam overlap with scanned and tilted (*left*) and overlapping focused beams (*right*)

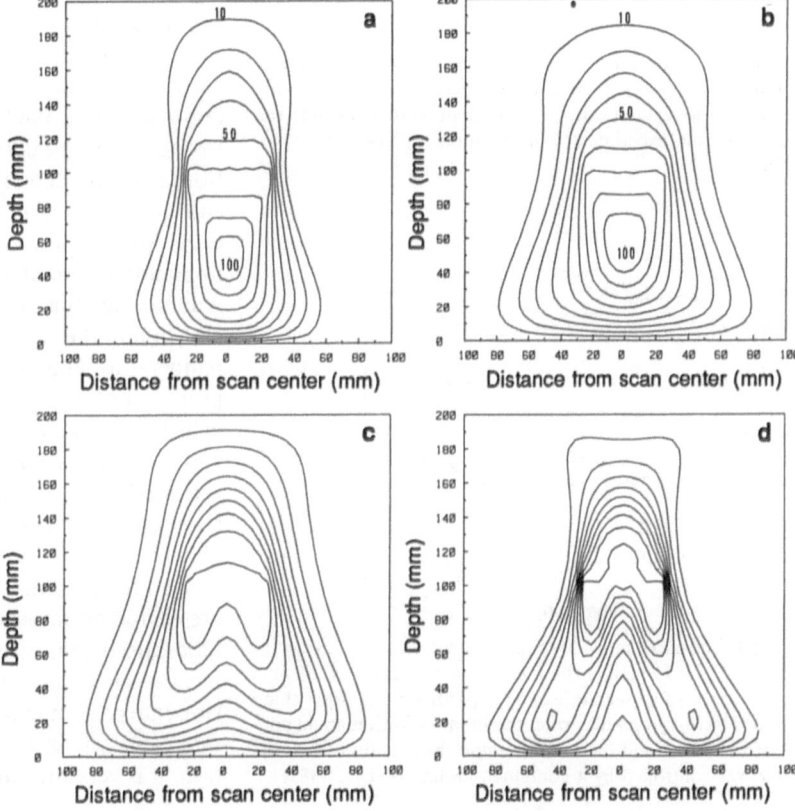

Fig. 2.52. The simulated steady-state temperature distributions produced by **a** a focused transducer (f-number = 1, frequency 1 MHz), **b** four focused and overlapping fields (f-number = 2), and **c, d** the same transducers except that they were sonicated in such a sequence that each beam was off when it crossed the central axis of the scan. The focal depth was 10 cm and the outer scan diameter was 50 mm. The perfusion rate was $0.5 \, \text{kg m}^{-3} \text{s}^{-1}$ in **a–c** and $5 \, \text{kg m}^{-3} \text{s}^{-1}$ in **d**. (Moros et al. 1989)

2.51, 2.52). In addition, some electric focusing techniques can be used to eliminate the beam overlapping problem by allowing destructive interference at the central axis of the transducer array (Cain and Umemura 1986).

2.6.4.3 Perfusion Effects

As with any hyperthermia method that utilizes energy deposition into the tumor, the temperature distributions induced with scanned focused ultrasonic beams are strongly perfusion dependent. Even in uniformly perfused tissues, both the magnitude and the distribution of the temperature varies with changes in perfusion. The temperature distribution appears to follow

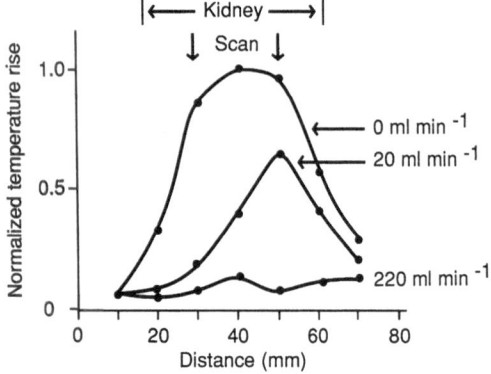

Fig. 2.53. The temperature distribution across a 20 mm diameter scan in a dog's kidney in vivo. The temperatures were measured after 10 min sonication with three different blood flows into the kidney (for the experimental information see Hynynen et al. 1987a)

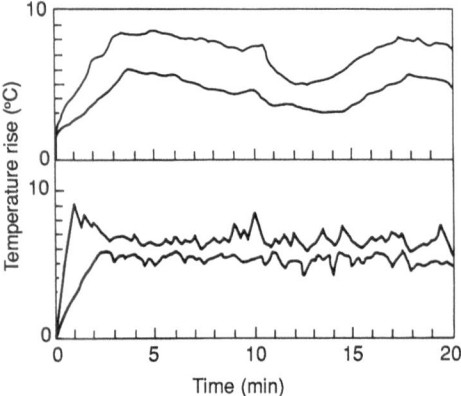

Fig. 2.54. The maximum and minimum temperature increases in a 30 mm diameter scan at the focal plane as a function of time during scanned focused ultrasound hyperthermia in dog's thigh in vivo. The *top graph* illustrates the situation with manual control of the power and the *bottom graph* the situation when the power was controlled by a computer. (Johnson et al. 1989)

the power deposition pattern more closely at high perfusion rates than at low perfusion rates, when the smoothing effects of thermal conduction are dominant. From in vivo experiments with a kidney, with a controllable perfusion rate (Hynynen et al. 1987a), it was shown that in certain situations when the blood perfusion and flow in the vessels (vessels oriented in the same direction) have a high degree of directionality, the temperature elevation can be smoothed by the presence of flow or perfusion (Fig. 2.53). These experiments also demonstrated the importance of the perfusion rate for the magnitude of the temperature elevation, showing that the temperature elevation measured in low perfusion case was 4–6 times higher than in kidney with normal perfusion. This makes it difficult to heat a well perfused tumor through a low perfusion tissue layer, even with strongly focused beams.

2.6.4.4 Feedback Control

It is generally known that the blood perfusion rate varies from tumor to tumor and also within a tumor (Jain and Ward-Hardley 1984). As was shown this can cause variations in the induced temperature distribution even with a uniform power deposition pattern. Thus, scanning the tumor with constant power does not necessarily induce uniform temperatures, and subtherapeutic temperatures have been measured in most of the large, deep tumors heated with scanned focused ultrasound without computerized feedback (Hynynen et al. 1987b). However, during the scanning the power can be controlled as a function of the location, and thus the amount and distribution of power can be modified as required to avoid hot spots and areas with too low a temperature. This can be done by implanting several thermocouples in the treatment volume, locating them in respect to the scan, and then controlling the power output based on the measured temperature and the location. This kind of a multipoint controller can significantly improve the achieved temperature distributions, as was demonstrated by Johnson et al. (1987) (Fig. 2.54), and such a controller has also been used in patient treatments (Das and Lele 1984). This approach of using completely decoupled single point controllers can lead to a good temperature distribution, provided that there are temperature sensors in all significantly different blood perfusion volumes. Sometimes this is impossible, especially since it is difficult to obtain the perfusion rate pattern in a clinical setting and the perfusion rate can vary as a function of temperature. In addition, it is very difficult to heat tissue volumes close to large blood vessels (diameter larger than 0.5 mm). Thus, there are tissue volumes that are almost certainly not properly heated, even during scanned focused hyperthermia utilizing a multipoint control system. However, multiregional feedback control does offer a significant improvement over the single point control algorithms.

2.6.4.5 Clincial Scanned Focused Ultrasound Hyperthermia Systems

In this section, a short description will be given of the two scanned focused ultrasound hyperthermia systems which are currently under clinical evaluation. Though neither of them is optimal for clinical treatments, the basic concepts are typical of mechanically scanned systems.

The MIT system developed by Dr. Lele and his co-workers was probably the first ultrasound system to be seriously considered for the induction of controlled hyperthermic temperatures in tumors. Since the first report (Lele 1975) this system has been improved and tested in over 200 patients. The mechanical scanning system has a robotic arm mounted on a x, y, z translator table on orthogonal axes. The arm can be tilted in two perpendicular planes to permit proper positioning of the sonication head. The transducers (up to four) are mounted on a

circular plate at the end of the arm. The plate rotates 360°, around and back, to allow high scanning speeds during the treatment. Three of the four transducers have to be mounted in a fixed position, and are thus able to execute only one scan radius each. One transducer can be moved radially and tilted by stepper motors. Thus, the system has eight degrees of motional freedom (Fig. 2.55), which allows complex scanning patterns to be executed. The ultrasound beams are generated by cross-cut quartz crystals (8, 12, or 16 cm in diameter) and focused by polystyrene lenses (focal length between 6 and 30 cm). The operating frequency can be selected to be between 0.6 and 6 MHz. The transducer head is coupled to the patient via an open plastic bag filled with degassed water. Acoustic gel is used between the plastic and the skin to aid in the transmission of ultrasound from the water into the tissue. The scanning pattern can be programmed and the sonication power can be controlled independently for each transducer by the operator or by the computer, with up to six independent PID controllers utilizing temperature measurements in the tumor. Thus, the system has the flexibility to treat tumors at different sites and also to control the power as a function of location in order to produce fairly uniform temperature distributions.

This system has been extensively tested with in vitro tissue samples and also in vivo, and it has been shown that scanned focused ultrasound can produce good

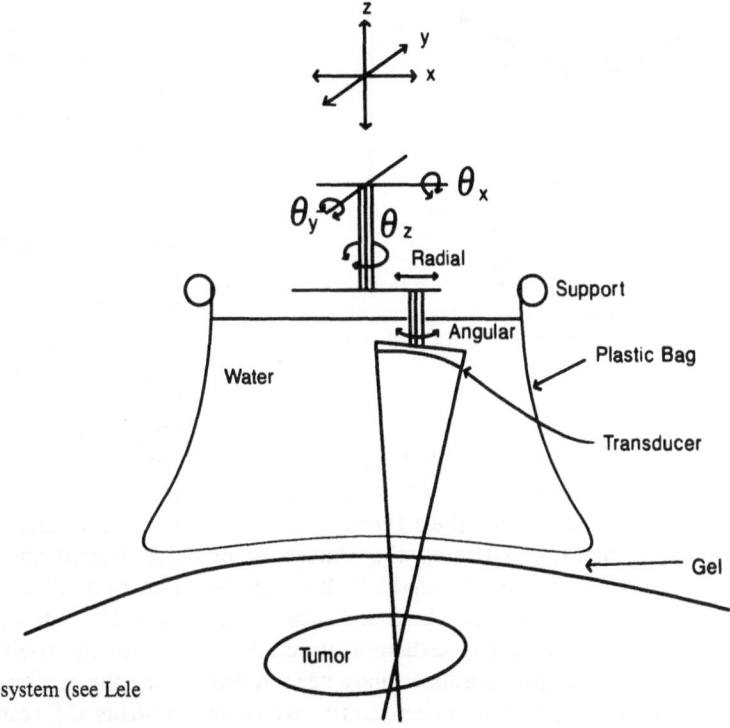

Fig. 2.55. The MIT scanned focused ultrasound system (see Lele 1983 for further details)

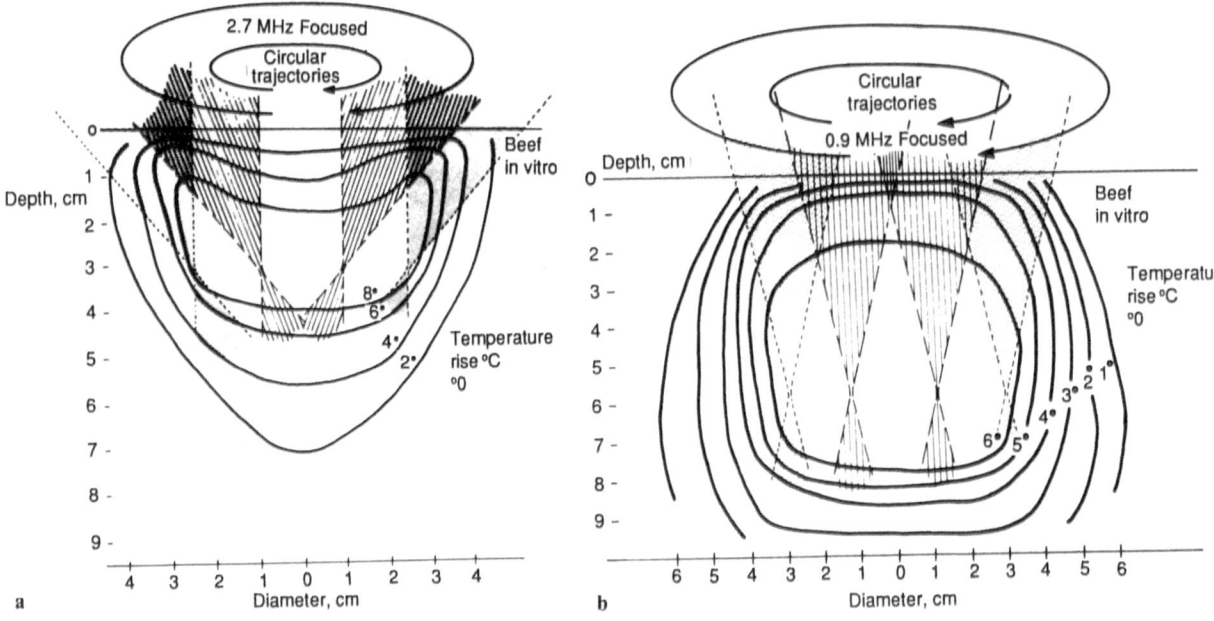

Fig. 2.56 a, b. The temperature distributions produced by the MIT system in beef in vitro. (Reprinted with permission; Lele and Parker 1982)

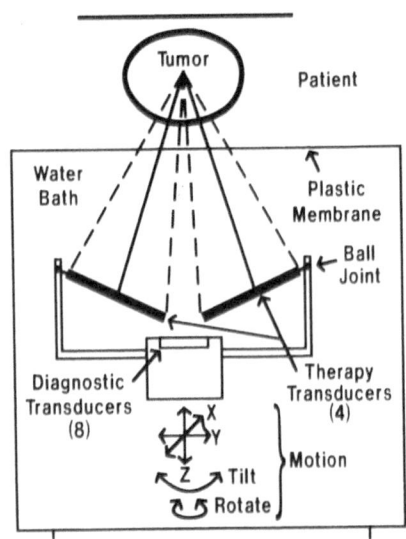

Fig. 2.57. The University of Arizona scanned focused ultrasound hyperthermia system (Hynynen et al. 1987a)

therapeutic temperature distributions (Fig. 2.56). Similar temperature distributions have been reported from the clinical treatments (Lele 1986).

The scanned focused system at the University of Arizona utilizes a commercial waterbath ultrasonic B-scanner (Octoson, Ausonics, Australia) which has been modified to become a hyperthermia device. A set of four focused therapy transducers have been added to the original imaging transducer gantry, which is immersed in a water bath. The foci of the four beams are overlapping in the imaging plane, although each beam can be oriented independently to point in other directions. The spherically curved transducers (diameter 130 mm and radius of curvature 250 mm) are custom built from ceramic transducer material and operate at a frequency of 1 MHz (designed for deep heating). Each transducer is independently driven from separate amplifiers; thus, any number of transducers can be on or at the desired power level. In addition, different transducers can be temporarily mounted on the side of the gantry for special treatments (for example brain treatments utilize a single focused 1.7-MHz transducer, the diameter and radius of curvature of which are 100 mm and 200 mm respectively). Each of the four transducers can also be replaced with different transducers when needed. The whole transducer gantry can be moved in five degrees of freedom (x, y, z, rotate, and tilt) by stepper motors (Fig. 2.57). A computer controls the motors and thus a desired scanning pattern can be programmed. The imaging capability of the system helps to direct the beams into the tumor and to avoid bone and gas. The tumor and other structures can be traced from the ultrasonic images (whenever the tumor is visible on the B-scan) and by using sequential scans a three-dimensional geometry of the treatment volume can be built into the computer memory. This information can then be used by the computer (or the operator) to plan the treatment. The system is also capable of utilizing CT scans to obtain the same information.

Table 2.7. Temperatures[a] obtained during scanned focused ultrasound treatments in the University of Arizona (Shimm et al. 1988)

Site	No. of treatments	Highest sensor	Lowest sensor	Sensors >42.5°C
Extremity (3 patients, 4 tumors)	10	44.7°C (41.5° – 49.2°C)	41.7°C (38.3° – 43.8°C)	35/78 (45%)
Head and neck (1 patient, 2 tumors)	3	42.0°C (41.6° – 42.3°C)	37.8°C (37.5° – 38.1°C)	0/68
Brain (2 patients, 2 tumors)	8	44.8°C (43.9° – 45.9°C)	41.8°C (41.0° – 42.9°C)	42/59 (71%)
Pelvis (8 patients, 9 tumors)	21	44.2°C (38.0° – 46.3°C)	40.6°C (38.0° – 43.1°)	110/199 (55%)

[a] Overall averages of the treatments are given, with the range in parentheses

The initial patient treatments have shown that this system can heat tumors at depth and that therapeutic temperatures are achievable in most of the measured locations (Table 2.7; Fig. 2.58). The immersion of the transducer gantry in the water bath is very convenient for heating tumors in certain locations, but is limiting when the patients cannot be positioned in such a manner that the beams can reach the tumor. In the case of abdominal and pelvic tumors, and tumors in chest wall and extremities, the patient lies on top of the water bath. Deep pelvis, rectal, and vaginal tumor patients sit on a specially made chair that has an opening in the seat, on top of the bath. In the brain treatments an individualized mould is used and a small part of the plastic membrane covering the water tank is boulused up to form a pathway for the ultrasound to propagate to the skin and through the opening in the skull bone into the tumor (Fig. 2.59).

The temperature distributions were found to be controllable during the treatment of small tumors (Fig. 2.60). In the case of large, deep tumors it was not possible to heat the whole tumor in one session; rather separate treatments had to be targeted to different locations of the tumor. In these initial treatments,

constant power was used during the scanning without computerized feedback control. Thus, these results are expected to represent the worst case situation and a significant improvement is expected with power modulated scanning (as has been shown by Lele 1983).

2.7 Technical Considerations in Ultrasound Hyperthermia

The first clinical patient treatments performed with nonfocused stationary fields were often power limited by pain, and thus the tumor temperatures did not reach a therapeutic level. The discomfort was generally present in treatments where bones were located close behind the tumor (Marmor et al. 1979; Corry et al. 1982), and was therefore linked to bone heating. These observations indicated the need for accurate treatment planning in order to avoid bone heating. However, when the basic physics of ultrasonic beam propagation through tissues is taken into account, it

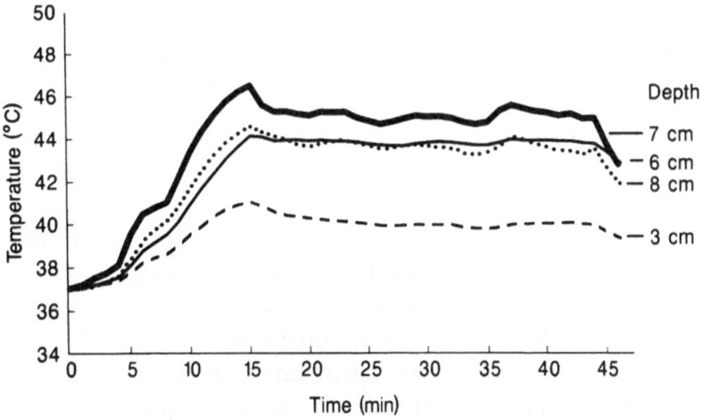

Fig. 2.58. Temperature as a function of time at various depths in a pelvic tumor during scanned focused ultrasound hyperthermia. The outer scan diameter was 5 cm and the focal depth, 7 cm. (Shimm et al. 1988)

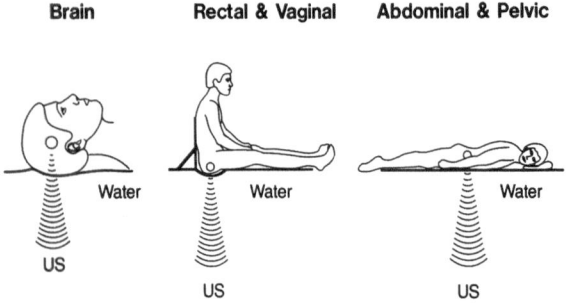

Fig. 2.59. The different treatment positions used with the University of Arizona ultrasound hyperthermia system

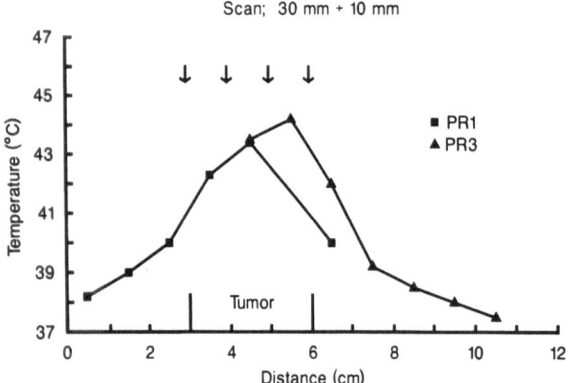

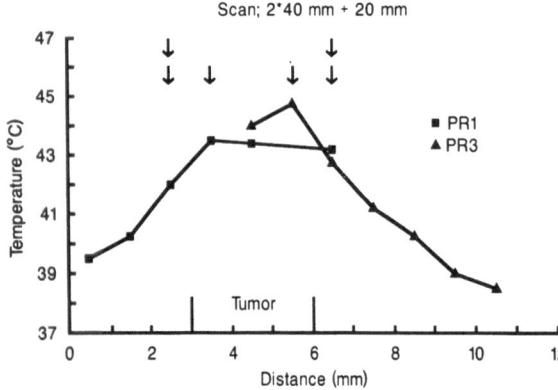

Fig. 2.60. The temperature across the scanned tumor volume in the vaginal wall at the depth of 3 cm during scanned focused ultrasound hyperthermia. *Top:* one external 3 cm and one internal 1 cm diameter scan. *Bottom:* two external 4 cm and one internal 2 cm diameter scans (Hynynen et al. 1987b). (PR1 and PR2 indicate temperatures measured with two separate seven-sensor thermocouple probes)

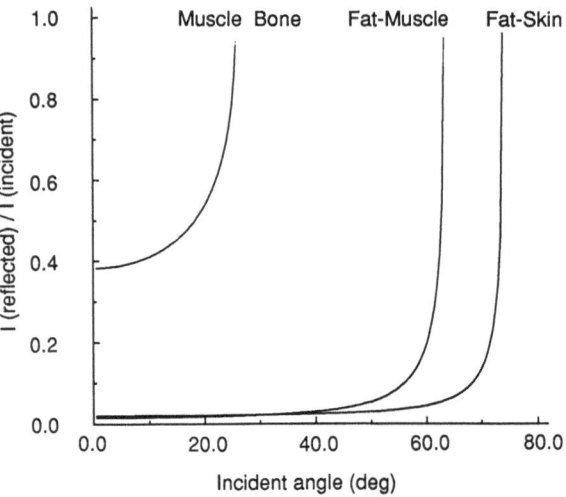

Fig. 2.61. The theoretical ratio between the reflected and incident beam at different tissue interfaces as a function of the incident angle

temperature distribution in the target volume, the tissue geometry and its relationship with respect to the temperature sensors and the heating field have to be known accurately. This results in a fairly complicated treatment procedure (when compared with superficial microwave or regional heating) and has probably been one of the factors discouraging the use of ultrasound hyperthermia.

2.7.1 Tissue Interfaces

From Table 2.1 (p. 78) it can be seen that there are only small variations in the acoustic properties of soft tissues. This means that the sound beam can penetrate well from one tissue layer into another without suffering large reflection losses. The situation is quite different at soft tissue-bone or -gas interfaces, where about 33% – 39% and 100% of the incident energy, respectively, is reflected back at normal incidence (Fig. 2.61). In addition, the amplitude attenuation coefficient of ultrasound is about 10–20 times higher in bone than in soft tissues. This causes the transmitted beam to be absorbed rapidly and thus a significant temperature increase results.

There are several studies showing preferential heating of the bone surface during sonication with nonfocused physiotherapy transducers operating at 1 MHz (Nelson et al. 1950; Lehmann et al. 1966, 1967). Similar hot spots also appear during ultrasound hyperthermia when weakly focused, low frequency, single beam systems are used to heat tissues in front of

is obvious that there are several other factors (such as tissue interfaces, gas, nonlinear propagation, the beam path available, and beam distortion) which need to be taken into account when deep tumors are heated using ultrasound. Thus, in order to obtain adequate

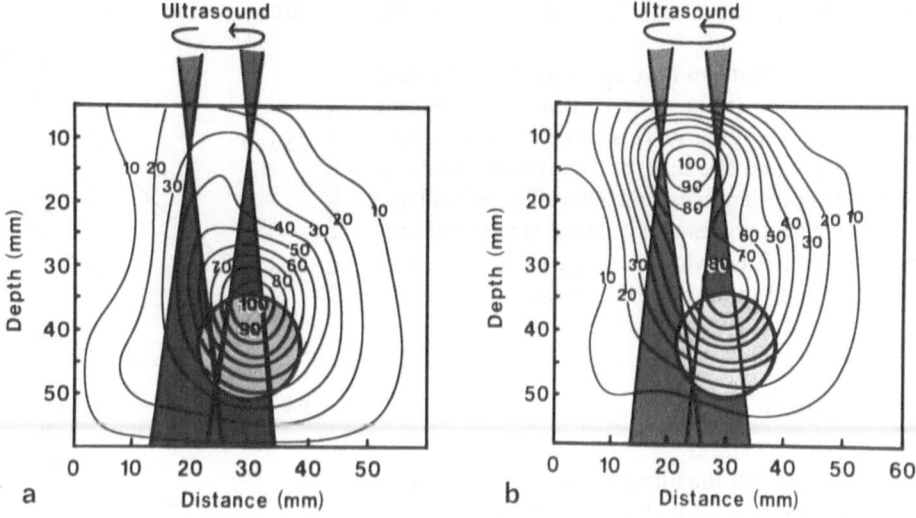

Fig. 2.62. Bone heating during scanned focused ultrasound hyperthermia. The measurements were done in dog's thigh in vivo and are plotted as normalized temperature elevation contours. The frequencies were 1 MHz (*left*) and 3.58 MHz (*right*) and the scan diameter was 10 mm. (Hynynen and DeYoung 1988)

bones (Marmor et al. 1979; Hynynen and DeYoung 1988) (Fig. 2.62). These hot spots can be avoided, or at least reduced to an acceptable level, if the intensities at the bone surface are between 10% (perfusion high) and 50% (perfusion low) of the value in the target volume. (This depends on the tumor perfusion rate, since well perfused tumors require more energy to obtain therapeutic temperatures.) The lower intensity at the bone surfaces can be achieved in many cases with multiple focused beams, higher frequencies, or more sharply focused transducers (Hynynen and DeYoung 1988). Since a hot spot is mainly caused by absorption in the bone, it can also be reduced by reducing the amount of energy transmitted into the bone through the interface. This can be effectively done by increasing the incident angle (Davis and Lele 1987).

The soft tissue-gas interfaces create a similar problem by reflecting all of the incident energy. This means that the absorbed power will be doubled close to the interface, and therefore the average power density of the beam has to be smaller than one-half of the value at the target volume (this again depends on the perfusion of the tumor and the tissues close to the interface). At lower frequencies the beams can propagate long distances in tissues; therefore, the gas interfaces can reflect the beams into unexpected locations and sometimes even cause a focal spot resulting in patient discomfort. Sometimes it is necessary to place plastic bags filled with water on the skin at the site where the beam will exit to allow the beam to propagate through the skin and out of the body. Acoustic gel can be used to increase the coupling between the bag and the skin.

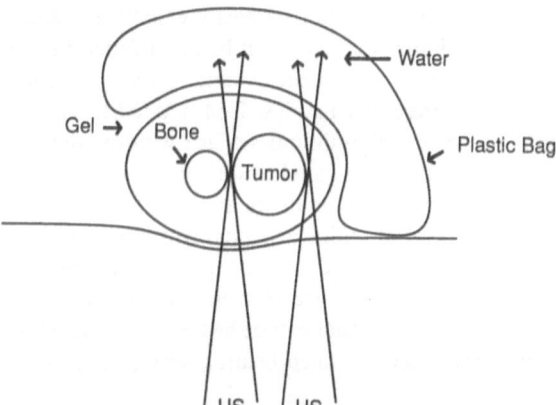

Fig. 2.63. The utilization of a water bag to prevent ultrasound reflections from a tissue-air interface

This technique has been found to reduce patient discomfort significantly in some of the ultrasound treatments of tumors (Fig. 2.63).

Because the acoustic impedances of different soft tissues are almost equal (Table 2.1), there is only a small amount of energy reflection when an ultrasonic beam propagates from one tissue to another. The largest difference is between fat and muscle tissues, which causes 1% − 2% of the beam to be reflected. When the incident angle is less than about 40° the amount of reflected energy remains fairly constant. At about 40° it starts to rise, reaching complete reflection at about 63.5° (Fig. 2.61). In most clinical situations the incident angles are smaller and the effect of the soft tissue interfaces can be ignored. However, when multiple tilted beams are used, the large incident angles can cause significant distortion in the heating field.

Another factor that can have an effect on the heating field distribution is the difference in the ultrasonic speed in different tissues. This can change the transmission angle (see Eq. 2.29) and thus alter the direction of the beam propagation. This effect is not large at small incident angles and becomes significant only when the beam is entering at a large angle and is propagating a long distance in the tissue.

2.7.2 Acoustic Window

Due to the fact that the ultrasonic beam is attenuated when it propagates through tissues, the beam has to enter through a large surface area and then converge through a smaller target area in order to induce the maximum temperature at the desired depth. The theoretical geometrical gain necessary to overcome the attenuation losses can be easily calculated, and the surface window diameter required as a function of depth for various frequencies and target volume diameters is presented in Figs. 2.64 and 2.65. These graphs indicate the minimum surface window diameter required to obtain equal absorbed power density at the surface and at depth. It appears that the required geometrical gain depends strongly on the operating frequency and also on the target volume diameter. The window size sets practical limitations on the size of tumors that can be heated at depth. The optimal frequency for deep heating seems to be be-

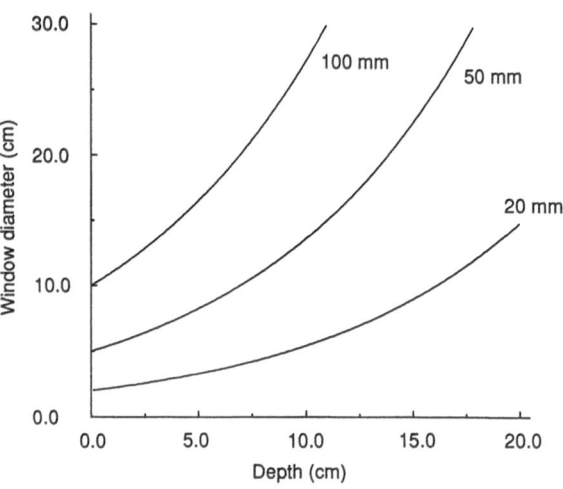

Fig. 2.65. The diameter of the ultrasonic window required to produce equal power absorption at the surface and at depth. The frequency is 1 MHz and the curves have been plotted for three different diameters of the target volume

tween 0.5 and 1 MHz, and in practice the highest possible frequency should be selected to minimize the hot spots at bone surfaces behind the tumor.

Another factor affecting the required geometrical gain and window size is the perfusion rate in the target volume and in the surrounding tissues. If the perfusion is lower in the tumor than in the overlying and underlying normal tissues, a smaller window size (than calculated based on the attenuation losses) will result in preferential target volume heating. However, if the perfusion is higher in the tumor, then a significantly larger window [up to double the diameter shown in Figs. 2.64 and 2.65 (Hynynen et al. 1987a)] may be required to obtain therapeutic temperatures in the tumor. Therefore, it is not always possible to heat a large, well perfused, and deep-lying tumor with even an ideal focused ultrasound system, as was shown by Fessenden et al. (1985). Technically, this problem can be solved by heating part of the tumor through the available ultrasonic window and then in the next treatment session another volume through the same window and so on until the whole tumor is treated. However, it is not yet known whether this approach results in a therapeutic response similar to that produced by a treatment in which the whole tumor is heated in the same session.

Brain tumors are good candidates for local therapy since the existing therapy fails locally. However, the skull bone surrounding the brain is almost impossible to penetrate with ultrasonic beams required for hyperthermia. In the past focused high intensity ultrasound was used to destroy brain tissues through an opening made in the skull (Fry 1965). A similar approach,

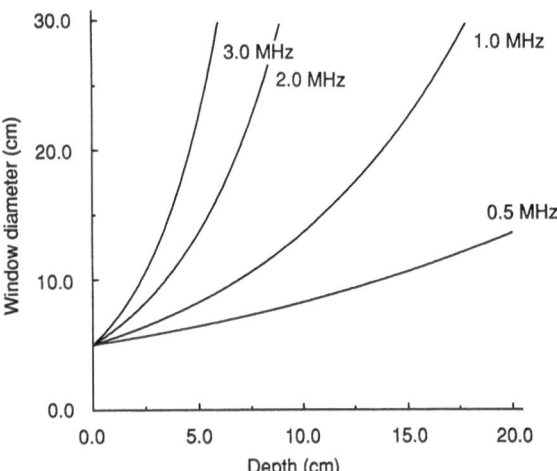

Fig. 2.64. The diameter of the ultrasonic window required for compensating the attenuation losses in soft tissues as a function of depth for various frequencies. The diameter of the target volume was 5 cm

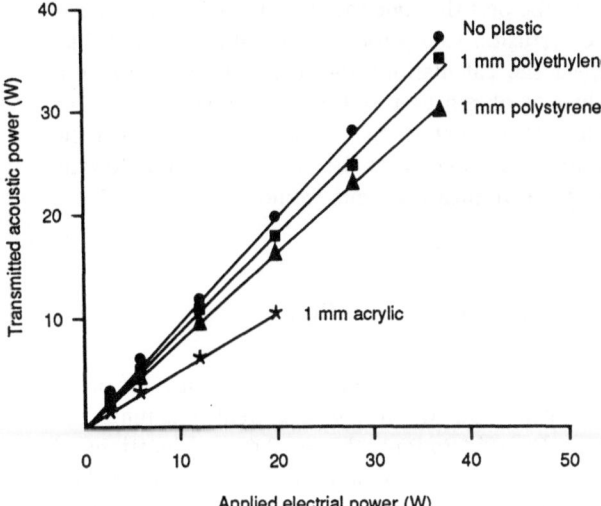

Fig. 2.66. The transmitted acoustic power through different plastic plates as a function of the applied electric power (Tobias et al. 1987)

where a piece of skull bone was removed and then the scalp replaced during a surgery, has successfully been used in initial clinical hyperthermia trials (Guthkelch et al. 1989). These experiments have shown that at least some brain tumors can be repeatedly treated and that good therapeutic temperatures can be reached in the tumor without heating the surrounding brain tissues. In the future, the skull bone may be replaced by an acoustic window made of plastic. It has been shown that ultrasound can be used to induce therapeutic temperatures in normal dog brains in vivo through a plate of polyethylene. Because the polyethylene window provides adequate mechanical protection, permanent replacement of the skull bone with the plate is possible (Fig. 2.66) (Tobias et al. 1987). Similarly, surgical creation of an ultrasonic window can probably be used in the treatment of other previously untreatable tumors.

2.7.3 Nonlinear Propagation

As was shown in Sects. 2.2 and 2.3, the propagation of an ultrasound wave is not linear at high power levels, but rather the wavefront will be distorted and higher harmonic frequencies will be generated. Since ultrasonic absorption is frequency dependent, the higher frequency components will increase power absorption from the field. For plane wave fields the nonlinearity becomes significant at intensities as low as $50\,\mathrm{W\,cm^{-2}}$ and $5\,\mathrm{W\,cm^{-2}}$ at frequencies of

2.2 MHz and 4.4 MHz, respectively (Carstensen et al. 1981). For focused 1-MHz fields significant nonlinear propagation (indicated by increased power absorption) has been found to start above $250\,\mathrm{W\,cm^{-2}}$ (Goss and Fry 1981). This means that at frequencies around 1 MHz and continuous wave sonication during ultrasound hyperthermia the propagation is almost always linear (this depends also on the beam shape). However, at higher frequencies and with weakly focused fields nonlinear propagation may be reached when large well perfused tumors (these require high ultrasound intensities) are treated.

Since the amount of wave distortion depends on the acoustic pressure, it has been proposed that this mechanism might be used to increase the amount of absorbed energy in the focal zone of a focused hyperthermia transducer where the highest acoustic pressures are achieved (Swindell 1985). This could be done by pulsing the ultrasonic field, thus maintaining constant power output, while increasing the temporal peak intensity of the beam to the range of the nonlinear propagation. Both theoretical (Swindell 1985) and experimental (Hynynen 1987) studies have shown that the increased power absorption is large enough to be useful during ultrasound hyperthermia. The absorbed power (and the temperature) increases more at the focus than in other parts of the field as a function of the intensity until the absorption in the overlying tissues starts to rise, thereby reducing the amount of energy reaching the focal zone (Fig. 2.67).

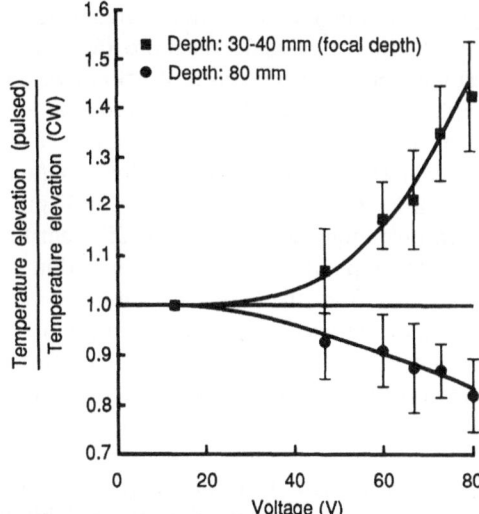

Fig. 2.67. The temperature elevation gain as a function of the driving voltage (intensity = voltage²) of the pulsed mode at the focus and about 40 mm beyond the focal depth. These measurements were done when a focused ultrasound transducer was scanned (scan diameter 10 mm) to heat the thigh muscle of a dog in vivo (Hynynen 1987)

In addition, the increased absorption at the focus reduces the amount of energy propagating to greater depths. Thus, nonlinear propagation can be utilized to improve the temperature distributions obtained during scanned focused ultrasound hyperthermia; this is desired especially when the tumor is deep and located close to a bone. However, despite all the promise, a considerable amount of work needs to be done before this phenomenon can be fully utilized during clinical treatments, and care should be taken not to elevate the intensity values above the threshold of transient cavitation in tissues.

2.7.4 Treatment Planning

Due to the highly localized nature of ultrasound hyperthermia and the complications created by bone and gas in the treatment field, the treatment geometry, i.e., the location of the target volume and the surrounding structures, has to be known accurately. The treatment technician has to be able to direct the ultrasonic beams into the target volume with the same precision as radiotherapy beams in order to cover the treatment volume properly, without overheating bones or surrounding normal tissues. Therefore, computerized, three-dimensional CT-based treatment planning of deep tumors is at least as important in ultrasound hyperthermia as it is in radiotherapy.

Hyperthermia treatment planning can be divided into two phases: (a) geometrical, in which the treatment geometry and the sonication parameters are optimized, and (b) thermal, which provides some kind of estimate of the temperature distribution. The most sophisticated treatment planning program to date has been described by Lele and Goddard (1987). In this program the anatomical information is obtained from sequential CT scans and then by utilizing average ultrasonic attenuation values for different tissue layers, the temporal average power deposition pattern is calculated from the assigned transducers and scanning path. The scanning path and angle together with the sonication parameters are then modified to give a power deposition pattern that covers the whole tumor and avoids bones. This may take several iterations before a desired result is obtained.

The thermal treatment planning programs do not have the same sophistication and reliability as the geometrical programs, mainly due to the difficulty of obtaining the perfusion values for different tissues and especially for tumors. In addition, the theoretical models of the cooling effect of blood perfusion have been criticized and not yet verified with experiments.

Thus, the best that one can do is to assign some average perfusion values for the normal tissues and then repeat the calculations for several tumor perfusion values to obtain best and worst case estimates. This will at least show when it is not possible to heat the tumor at all, and what the critical locations for the temperature measurements would be.

2.7.5 Treatment Execution

The first step in a hyperthermia treatment is the probe placement. Based on the treatment planning and clinical considerations, the number and locations of multisensor thermocouples is decided. Then the clinician inserts the temperature sensors into the tumors. Probe placement is most accurate if it is done under CT or ultrasound guidance. If this is not possible then probe location should be verified by CT scan or orthogonal X-rays before the treatment in orders to be able to control the ultrasound power based on the measured temperatures during the sonication.

Following probe placement the patient is positioned for the treatment. Since even the slightest movement (a few millimeters) can cause the beam to hit bone (especially when a tumor close to a bone is treated), patient movements have to be eliminated as much as possible. This can be done by using individualized molds, belts, sandbags, etc. In order to be able to obtain the treatment geometry and to direct the treatment field accurately, an ultrasonic imaging transducer can be utilized in the heating transducer head (Hynynen et al. 1987 a, b). The ultrasound image can be used to ascertain the location of the skin, bones, and tumor with respect to the heating field. In addition, if the thermocouples have been inserted in such a manner that they are not parallel with the imaging beam, then even the temperature sensing probes can be seen (Fig. 2.68). However, it is not always possible to visualize the tumor and the probes. The sensor locations also can be located by scanning a focused ultrasound beam along a raster pattern while simultaneously monitoring the temperatures of each sensor. When the focus passes the thermocouple location a sharp temperature elevation is detected and thus the location of each sensor can be obtained. In a case in which the tumor cannot be visualized, the thermocouple locations and the geometrical information can be used to direct the beams into the tumor.

During the treatment the power is increased gradually and if patient discomfort is detected before reaching therapeutic temperatures, a modification of the power deposition pattern can be made (i.e., a slight change

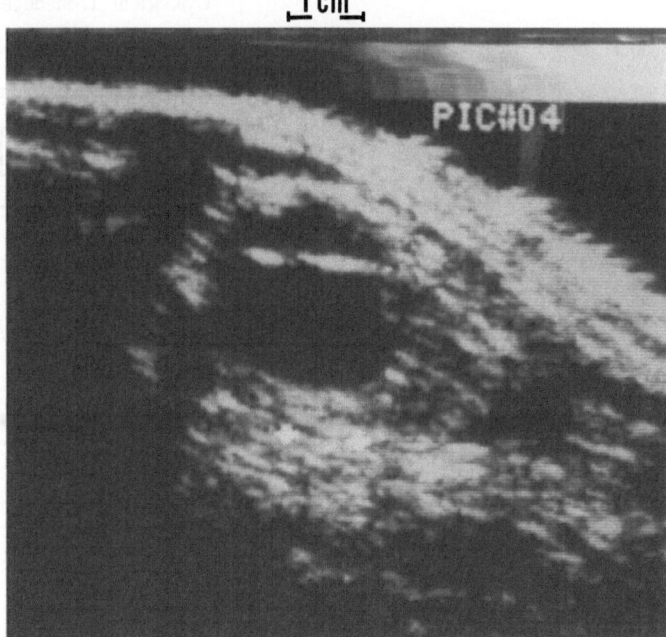

Fig. 2.68. An ultrasound image of a superficial human tumor. Note the thermocouple probe passing through the tumor. The image was obtained with the modified Octoson scanner described by Hynynen et al. (1987a)

in the scan location, size, depth, or beam angle, and power). Often even a small modification can help to relieve the pain. At present it appears that the utilization of the temperature information from previous treatments significantly helps to improve the achieved temperature distributions.

Although ultrasound hyperthermia treatments may sound very complex, they offer a significant improvement over other noninvasive heating techniques. Thus, if one desires to achieve an adequate hyperthermia treatment depth, focused ultrasound is the method of choice. It may be that in the future most of the above-mentioned tasks (i.e., thermocouple location, heating field optimization, etc.) can be done automatically by the heating system, making the treatments less time consuming and easier to execute.

2.8 Future Developments in Ultrasound Hyperthermia

So far, only some of the potential of ultrasound as a method of inducing elevated temperatures has been utilized, and the systems developed have been far from optimal. However, the possibilities have been demonstrated and thus clinically useful systems need to be developed and manufactured. The treatment planning and control programs need to be improved to the same level as the treatment planning software for radiation therapy. It is likely that there will be further development of scanned focused ultrasound systems. The mechanical scanning presently used will be utilized for a few years, but eventually the electrically scanned and focused systems will become available in the clinic, and the electrical systems may finally become more popular than the mechanical devices. In addition, some combination of mechanical and electrical scanning and focusing may prove useful. Special applicators need to be developed to treat tumors at sites that cannot be heated with the general purpose devices. Finally, the optimal treatment, i.e., temperature, exposure time, treatment volume, number of treatments, interval between treatments, etc., is not yet known. Therefore, it appears that both commercial development and technical and clinical research face many challenges before the optimal way of using ultrasound hyperthermia will be established. In the following only three of the areas under strong development will be discussed.

2.8.1 Control

Automatic power control, and thus temperature control in the tumor and surrounding tissues, has great potential for improving the achieved temperature distributions and therefore the clinical response to hyperthermia treatments (Roemer et al. 1986). These control algorithms will be developed with new systems

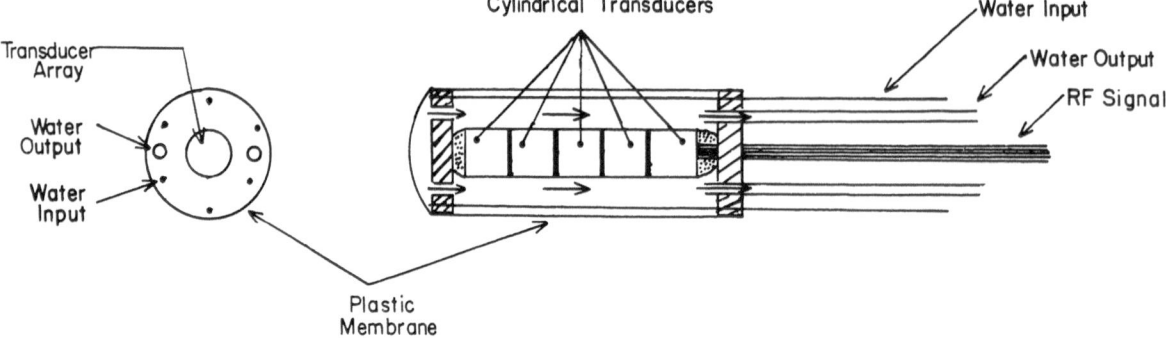

Fig. 2.69. A diagram of the experimental intracavitary ultrasonic applicator (Diederich and Hynynen 1987)

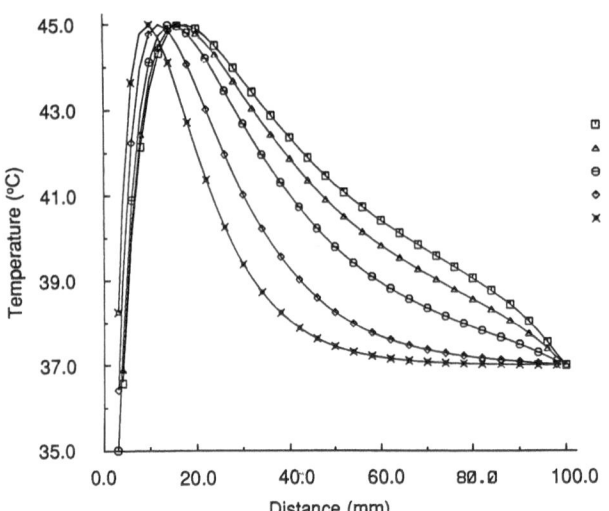

Fig. 2.70. Theoretical radial temperature distributions induced by a cylindrical intracavitary applicator (diameter 10 mm, surface temperature 22 °C) as a function of the operating frequency (Diederich and Hynynen 1987)

and thus there should be a significant improvement in the temperature homogeneity achieved during treatments. In addition, it may well be that estimation techniques could be used to obtain the complete temperature distribution even between the temperature measurement points, and that these estimated temperature values could then be used to control the whole power field (Kress 1987). However, these techniques require a considerable amount of computing power and it will take several years before they can be employed in clinical treatments.

2.8.2 Special Applicators

If a tumor is located close to a body cavity, an intracavitary energy source can be used. The first use of intracavitary hyperthermia was recorded in 1898, when hot water circulating through copper rods was utilized to treat cervical cancer (Westermark 1898). To date, most intracavitary applicators using microwaves have been employed to treat esophageal, vaginal, and rectal tumors in addition to cancer of the prostate (such treatment of tracheal tumors has also been proposed). Although microwaves give better heating at depth than the hot water heater, the power deposition pattern is not easily controllable. In theory, intracavitary ultrasonic applicators with multiple transducer elements appear to offer good control over the deposited power. A simple prototype device constructed from separate cylindrical transducer elements, each of which can be driven individually, is shown in Fig. 2.69. The computer simulations showing the potential of this approach (Fig. 2.70) are very promising. Also, the preliminary in vivo experiments have shown that the applicator can control the resulting temperature distribution (Diederich and Hynynen 1987, 1989a). In the future, electrically focused and scanned systems will probably improve the control over the power deposition pattern from an intracavitary applicator and thus make it possible to heat even deeper tumors (Diederich and Hynynen 1989b). This approach of using intracavitary ultrasound in the induction of hyperthermia offers a lot of promise, since the system could be made easy to operate and thus the treatment could be adapted as a routine therapy with simple and accurate delivery of energy directly into the tumor.

Similar special applicators will probably be designed to heat different tumors at various anatomical locations to improve the temperature distributions obtained during hyperthermia treatments.

2.8.3 High Temperature Hyperthermia

During clinical hyperthermia treatments, significant temperature variations are caused by spatial and temporal variations in blood perfusion rate. Thus, temperatures must be mapped throughout the tumor and surrounding normal tissues in order to control the power deposition pattern adequately during sonication. In a clinical treatment, the number of invasive temperature measurement probes is limited, and thus generally there is not enough information to produce uniform, or even nonuniform but therapeutic, temperatures in the whole tumor volume. A possible way to overcome the effect of perfusion is to perform the treatment in a short period of time at a high enough temperature to obtain a therapeutic effect. Independence of temperature elevation from blood perfusion rate during the very beginning of the sonication has been shown by simulations and using an in vitro perfused kidney model (Billard et al. 1988) (Fig. 2.71). Thus, if the local absorbed power could be estimated, it would be possible to calculate the temperature elevation and the thermal exposure in that loca-

tion. This approach has potential because the tissue density, specific heat, and thermal conduction (the factors determining the temperature elevation during short periods of energy input) are fairly well known for tissues and do not vary to the same extent as blood perfusion rate (Bowman 1982).

In principle this method was first proposed (though for incorrect reason) by Burov (1956), as was discussed in Sect. 2.1. He used intensities of 150 W cm^{-2} at 1.5 MHz for 1–3 s, which would in theory give a perfusion-independent temperature increase somewhere between 50° and 60°C for the 3-s pulse. This can be a therapeutically significant thermal exposure and probably explains the promising results obtained using this sonication pattern both in animal and in human tumors (Burow 1956; Burov and Andreevskaya 1956; Oka 1960). More recently, the short sound pulses have been used at levels which produce coagulation necrosis in the treatment of animal tumors (Fry and Johnson 1978), and were proposed by Britt et al. (1984) for delivering an accurate thermal dose during hyperthermia treatments of the brain. However, high temperature hyperthermia is still far away from being used in clinical treatments of deep tumors because systematic studies of the various parameters affecting the temperature increase have yet to be done. Eventually, with the aid of a few implanted thermocouples to calibrate the sonication system for each tumor, this technique may be used to improve the thermal dose delivered to the whole tumor.

References

Apfel RE (1986) Prediction of tissue composition from ultrasonic measurements and mixture rules. J Acoust Soc Am 79:148–152

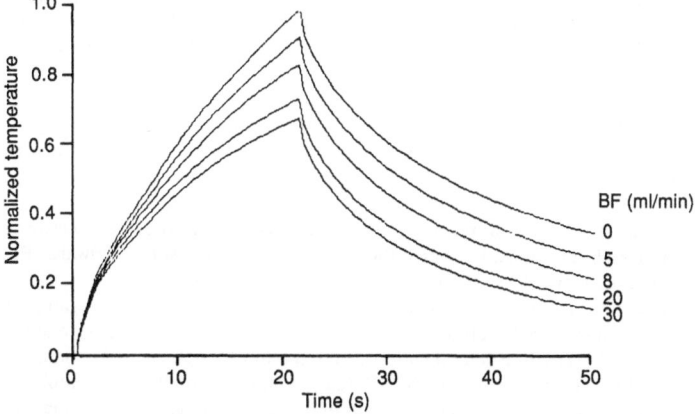

Fig. 2.71. The normalized temperature rise during a 20-s sound pulse (1 MHz, focused transducer, diameter 120 mm, radius of curvature 250 mm), measured with an uncoated thermocouple (wire diameter 50 μm) in a fixed dog kidney (Holmes et al. 1984) with different flows of water into the kidney (Billard et al. 1988)

Bamber JC, Hill CR (1979) Ultrasonic attenuation and propagation speed in mammalian tissues as a function of temperature. Ultrasound Med Biol 5:149–157

Bamber JC, Nassiri DK (1985) Effect of gaseous inclusions on the frequency dependence of ultrasonic attenuation in liver. Ultrasound Med Biol 11:293–298

Beard RE, Magin RL, Frizzell LA, Cain CA (1982) An annular focus ultrasonic lens for local hyperthermia treatment of small tumors. Ultrasound Med Biol 8:177–184

Benkeser PJ, Frizzell LA, Ocheltree KB, Cain CA (1987) A tapered phased array ultrasound transducer for hyperthermia treatment. IEEE Trans Ultrason Ferroelectr Freq Control UFFC 34:446–453

Billard B, Hynynen K, Roemer RB (1988) Induction of perfusion independent thermal exposure. Proc 5th Int Symp Hyperthermic Oncology, Kyoto, Japan, pp 713–714

Bjorno L (1986) Characterization of biological media by means of their nonlinearity. Ultrasonics 24:254–259

Britt RH, Pounds DW, Lyons BE (1984) Feasibility of treating malignant brain tumors with focused ultrasound. Prog Exp Tumor Res 28:232–245

Bowman FH (1982) Heat transfer mechanisms and thermal dosimetry. Natl Cancer Inst Monogr 61:437–445

Burov AK (1956) High intensity ultrasonic oscillation for the treatment of malignant tumors in animals and man. Dokl Akad Nauk SSSR 106:239–241

Burov AK, Andreevskaya GD (1956) The effect of ultra-acoustic oscillation of high intensity on malignant tumors in animals and man. Dokl Akad Nauk SSSR 106:445–448

Cain CA, Umemura S-A (1986) Concentric-ring and sector vortex phased array applicators for ultrasound hyperthermia therapy. IEEE Trans Microwave Theory Tech MTT 34:542–551

Cain CA, Umemura S, Ibbini M, Ebbini E (1987) Ultrasound phased array hyperthermia applicators. In: Proceedings of the 9th IEEE Engineering in Medicine and Biology Society meeting, pp 1640–1641. IEEE Catalog No 87CH2513-0

Calderon C, Vilkomerson D, Mezrich R, Etzold KF, Kingsley B, Haskin M (1976) Differences in the attenuation of ultrasound by normal, benign, and malignant breast tissue. J Clin Ultrasound 4:249–254

Carstensen EL (1987) Acoustic cavitation and the safety of diagnostic ultrasound. Ultrasound Med Biol 13:597–606

Carstensen EL, Muir TG (1986) The role of nonlinear acoustics in biomedical ultrasound. In: Greenleaf JF (ed) Tissue characterization with ultrasound, vol 1. CRC, Boca Raton, pp 57–79

Chan AK, Sigelmann RA, Guy AW (1974) Calculations of therapeutic heat generated by ultrasound in fat-muscle-bone layers. IEEE Trans Biomed Eng BME 21:280–284

Chin RB, Madsen EL, Zagzebski JA, Frank GR (1986) Experimental tests of computed time-dependent temperature distributions during ultrasonic heating. Program and abstracts of IEEE 1986 Ultrasonic Symp, p 150

Chivers RC, Parry RJ (1978) Ultrasonic velocity and attenuation in mammalian tissues. J Acoust Soc Am 63:940–953

Clarke PR, Hill CR, Adams K (1970) Synergism between ultrasound and x-rays in tumor therapy. Br J Radiol 43:97–99

Coleman DJ, Lizzi FL, Burgess SEP, Silverman RH, Smith ME, Driller J, Rosado A, Ellsworth RM, Haik BG, Abrahamson DH, McCornick B (1986) Ultrasonic hyperthermia and radiation in the management of intraocular malignant melanoma. Am J Ophthalmol 101:635–642

Cook BD (1977) Ultrasonic radiation determination by optical methods. In: Hazzard DG, Litz ML (eds) Symposium on biological effects and characterizations of ultrasound sources. HEW publication (FDA) 78-8048. United States Department of Health, Education and Welfare, Rockville, MD, pp 99–105

Corry PM, Barlogie B, Tilchen EJ, Armour EP (1982) Ultrasound induced hyperthermia for the treatment of human superficial tumors. Int J Radiat Oncol Biol Phys 8:1225–1229

Coughlin CT, Colacchio T, Crichlow R, Ryan T, Strohbehn J (1987) Ultrasound-induced intraoperative hyperthermia. Proceedings of the 35th annual meeting of the Radiation Research Society, Atlanta, p 16

Das H, Lele PP (1984) Design of a power modulator for control of tumor temperature. In: Overgaard J (ed) Hyperthermic oncology 1984, vol 1. Taylor and Francis, London, pp 707–714

Davis BJ, Lele PP (1987) Bone-pain during hyperthermia by ultrasound. Proceedings of the 35th annual meeting of the Radiation Research Society, Atlanta, p 11

Diederich C, Hynynen K (1987) Induction of hyperthermia using an intracavitary ultrasonic applicator. Proceedings of the IEEE Ultrasonic Symp IEEE Catalog No 87CH2492-7, pp 871–874

Diederich C, Hynynen K (1989a) Induction of hyperthermia using an intracavitary multi-element ultrasonic applicator. IEEE Trans Biomedical Engineering 36:432–438

Diederich C, Hynynen K (1989b) The feasibility of intracavitary ultrasound hyperthermia using an electrically focussed multielement array. Proc 37th annual meeting of the Radiation Research Society and 9th annual meeting of the North American Hyperthermia Group, Seattle, Washington, p 9

Do-Huun JP, Hartemann P (1982) Deep and local heating induced by an ultrasound phased array transducer. Proceedings of the IEEE Ultrasonic Symp 735–738

Dunn F (1976) Ultrasonic attenuation, absorption, and velocity in tissues and organs. In: Linzer M (ed) Ultrasonic tissue characterization. NBS special publication no 453. NBS, Washington, pp 21–28

Dunn F, Averbuch J, O'Brien WD (1977) A primary method for determination of ultrasonic intensity with elastic sphere radiometer. Acoustica 38:58–61

Dunn F, Brady JK (1973) Absorption of ultrasound in biological media. Biophysics 18:1128–1132

Dyer HJ (1972) Structural effect of ultrasound on the cell. In: Reid JM, Sikov MR (eds) Interaction of ultrasound and biological tissues. DHEW publication (FDH) 73-8008. United States Department of Health, Education and Welfare, Rockville, MD, pp 73–75

Dyson M, Pond JB, Woodward B, Broadbent J (1974) The production of blood cell stasis and endothelial damage in the blood vessels of chick embryos treated with ultrasound in a standing wave field. Ultrasound Med Biol 1:133–148

Ebbini ES, Umemura S-I, Ibbini M, Cain C (1988) A cylindrical-section ultrasound phased array applicator for hyperthermia cancer therapy. IEEE Trans Ultrason Ferroelectr Freq Control 35:561–572

Edmonds PD, Ross WC, Lee ER, Fessenden P (1985) Spatial distributions of heating by ultrasound transducers in clinical use, indicated in a tissue-equivalent phantom. Proceedings of the IEEE Ultrasonic Symp, pp 908–912

Edwards PL, Jarzynski J (1980) Use of a microsphere probe for pressure field measurements in the megahertz frequency range. J Acoust Soc Am 68:356–359

Fessenden P, Lee ER, Anderson TL, Strohbehn JW, Meyer JL, Samulski TV, Marmor JR (1984) Experience with a multitransducer ultrasound system for localized hyperthermia of deep tissues. IEEE Trans Biomed Eng BME 31:126–135

Fessenden P, Meyer JL, Valdagni R, Lee ER, Samulski TV, Kapp DS, Bagshaw MA (1985) Analysis of deep hyperthermia treatments using six ultrasound transducers in a fixed frequency/fixed geometry configuration. In: Proceedings of the Annual Meeting of the Radiation Research Society, Los Angeles, California, p 15

Foster FS, Hunt JW (1980) The focussing of ultrasound beams through human tissue. Acoust Imaging 8:709–718

Frizzell LA, Carstensen E (1976) Shear properties of mammalian tissues at low megahertz frequencies. J Acoust Soc Am 60:1409–1411

Frizzell LA, Lee CS, Aschenbach PD, Borrelli MJ, Morimoto RS, Dunn F (1983) Involvement of ultrasonically induced cavitation in the production of hind limb paralysis of the mouse neonate. J Acoust Soc Am 74:1062–1065

Frizzell LA, Miller DL, Nyborg WL (1986) Ultrasonically induced intravascular streaming and thrombus formation adjacent to a micropipette. Ultrasound Med Biol 12:217–221

Fry FJ (1965) Recent developments in ultrasound at biophysical research laboratory and their application to basic problems in biology and medicine. In: Kelly E (ed) Ultrasound energy. University of Illinois Press, Urbana, pp 202–228

Fry WJ, Dunn F (1962) Ultrasound: analysis and experimental methods in biological research. In: Nastuk WM (ed) Physical techniques in biological research, vol 4. Special methods. Academic, New York, pp 261–325

Fry WJ, Fry RB (1954a) Determination of absolute sound levels and acoustic absorption coefficients by thermocouple probes – theory. J Acoust Soc Am 26:294–310

Fry WJ, Fry RB (1954b) Determination of absolute sound levels and acoustic absorption coefficients by thermocouple probes – experiments. J Acoust Soc Am 26:311–317

Fry FJ, Johnson LK (1978) Tumor irradiation with intense ultrasound. Ultrasound Med Biol 4:337–341

Goss SA, Fry FJ (1981) Nonlinear acoustic behavior in focussed ultrasonic fields: observations of intensity dependent absorption in biological tissue. IEEE Trans Sonics Ultrason SU 28:21–26

Goss SA, Johnson RL, Dunn F (1978) Comprehensive compilation of empirical ultrasonic properties of mammalian tissues. J Acoust Soc Am 64:423–457

Goss SA, Frizzell LA, Dunn F (1979) Ultrasonic absorption and attenuation of high frequency sound in mammalian tissues. Ultrasound Med Biol 5:181–186

Goss SA, Johnson RL, Dunn F (1980) Compilation of empirical ultrasonic properties of mammalian tissues. II. J Acoust Soc Am 68:93–108

Guthkelch AN, Hynynen K, Shimm D, Stea B, Cassady JR, Roemer RB (1989) Treatment of malignant brain tumors with focussed ultrasound hyperthermia and radiation: experiences with a phase I trial. J Neurosurgery (submitted)

Hahn GM (1982) Does the mode of heat induction modify drug anti-tumor effects? Br J Cancer 45 (Suppl V):238–242

Haran ME (1977) Ultrasonic acousto-optic measurement techniques. In: Symposium on biological effects and characterization of ultrasound sources. HEW publication (FDA) 78-8044. United States Department of Health, Education, and Welfare, Rockville, MD, Hazzard DG, Litz ML (eds) pp 90–98

Heimburger RF (1985) Ultrasound augmentation of central nervous system tumor therapy. Indiana Med 78:469–476

Hill CR (1972) Ultrasonic exposure thresholds for changes in cells and tissues. J Acoust Soc Am 52:667–672

Holmes KR, Ryan W, Weinstein P, Chen MM (1984) A fixation technique for organs to be used as perfused tissue phantoms in bioheat transfer studies. 1984 Advances in Bioengineering, Spiker RL (ed) (New York: American Society of Mechanical Engineers), 9–10

Horvath J (1944) Ultraschallwirkung beim menschlichen Sarkom. Strahlentherapie 75:119

Hueter TF, Bolt RH (1955) Sonics; techniques for the use of sound and ultrasound in engineering and science. Wiley, New York

Hunt JW (1985) Review of deep heating using ultrasonic beams. Proceedings of the 33th annual meeting of the Radiation Research Society, Los Angeles, California, p 16

Hynynen K (1987) Demonstration of enhanced temperature elevation due to nonlinear propagation of focussed ultrasound in dog's thigh in vivo. Ultrasound Med Biol 13:85–91

Hynynen K, DeYoung D (1988) Temperature elevation at muscle-bone interface during scanned, focussed ultrasound hyperthermia. Int J Hyperthermia 4:267–279

Hynynen K, Watmough DJ, Mallard JR (1981) Design of ultrasonic transducers for local hyperthermia. Ultrasound Med Biol 7:397–402

Hynynen K, Watmough DJ, Mallard JR, Fuller M (1983a) The construction and assessment of lenses for local treatment of malignant tumors by ultrasound. Ultrasound Med Biol 9:33–38

Hynynen K, Watmough DJ, Shammari M, Wilmot G, Murthy MSN, Mallard JR, Fuller M, Sarkar T (1983b) A clinical hyperthermia unit utilizing an array of seven focussed ultrasonic transducers. In: Proceedings of the IEEE Ultrasonic Symp, pp 816–821

Hynynen K, Roemer R, Moros E, Johnson C, Anhalt D (1986) The effect of scanning speed on temperature and equivalent thermal exposure distributions during ultrasound hyperthermia in vivo. IEEE Trans Microwave Theory Tech MTT 34:552–559

Hynynen K, Roemer R, Anhalt D, Johnson C, Xu ZX, Swindell W, Cetas TC (1987a) A scanned, focussed, multiple transducer ultrasonic system for localized hyperthermia treatments. Int J Hyperthermia 3:21–35

Hynynen K, Shimm D, Roemer RB, Anhalt D, Cassady JR (1987b) Temperature distributions during clinical ultrasound hyperthermia. Proceedings of the 9th annual conference of IEEE Engineering in Medicine and Biology Society, Boston, Nov 1987. IEEE, New York, pp 1644–1645

Jain RK, Ward-Hardley K (1984) Tumor blood flow – characterization, modification and role in hyperthermia. IEEE Trans Sonics Ultrason SU 31:504–526

Johnson C, Kress R, Roemer RB, Hynynen K (1987) Multipoint feedback control system for scanned, focussed ultrasound hyperthermia. Proc 35th annual meeting of Radiation Research Society, Atlanta, Georgia, p 12

Johnson SA, Christensen DA, Johnson CC, Greenleaf JF, Rajagopalan B (1977) Non-intrusive measurement of microwave and ultrasound induced hyperthermia by acoustic temperature tomography. Proceedings of the IEEE Ultrasonic Symp, pp 977–982

Ibbini MS, Cain CA (1989) A field conjugation method for direct synthesis of hyperthermia phased array heating patterns. IEEE Trans Ultrason Ferroelectr Freq Control 36:3–9

Kikuchi Y, Uchida R, Tanaka K, Wagai T (1957) Early diagnosis through ultrasonics. J Acoust Soc Am 29:824–833

Kishi M, Mishima T, Itakura T, Tsuda K, Oka M (1975) Experimental studies of effects of intense ultrasound on implantable murine glioma. In: Proceedings of the 2nd European congress on ultrasonics in medicine. Exerpta Medica, Amsterdam, pp 28–33

Kossoff G (1979) Analysis of focusing action of spherically curved transducers. Ultrasound Med Biol 5:359–365

Kremkau FW (1979) Cancer therapy with ultrasound: a historical review. J Clin Ultrasound 7:287–300

Kress RL (1987) Adaptive model – following control for hyperthermia treatment systems. Ph D thesis, Department of Aerospace and Mechanical Engineering, University of Arizona

Law WK, Frizzell LA, Dunn F (1985) Determination of the nonlinearity parameter B/A of biological media. Ultrasound Med Biol 11:307–318

Lehmann JF (1965) Ultrasound therapy. In: Licht S (ed) Therapeutic heat and cold. Licht, New Haven, pp 321–386

Lehmann JF, deLateur BJ, Silverman DR (1966) Selective heating effects of ultrasound in human beings. Arch Phys Med Rehabil 47:331–339

Lehmann JF, deLateur BJ, Warren CG, Stonebridge JS (1967) Heating produced by ultrasound in bone and soft tissue. Arch Phys Med Rehabil 48:397–401

Lele PP (1975) Hyperthermia by ultrasound. In: Proceedings of the international symposium on cancer therapy by hyperthermia and radiation. Washington, April 28–30, pp 168–178

Lele PP (1977) Thresholds and mechanisms of ultrasonic damage to "organized" animal tissues. In: Hazzard DG, Litz ML (eds) Symposium on biological effects and characterizations of ultrasound sources. DHEW publication FDA 78-8048. United States Department of Health, Education and Welfare, Rockville, MD, pp 224–239

Lele PP (1981) An annular-focus ultrasonic lens for production of uniform hyperthermia in cancer therapy. Ultrasound Med Biol 7:191–193

Lele PP (1983) Physical aspects and clinical studies with ultrasound hyperthermia. In: Storm FC (ed) Hyperthermia in cancer therapy. Hall Medical, Boston, pp 333–367

Lele PP (1984) Ultrasound: is it the modality of choice for controlled, localized heating of deep tumors? In: Overgaard J (ed) Hyperthermic oncology 1984, vol 2. Taylor and Francis, London, pp 129–154

Lele PP (1986) Rationale, technique and clinical results with scanned focussed ultrasound (SIMFU) systems. Proceedings of the 8th annual conference of IEEE Engineering in Medicine and Biology Society. IEEE, New York, pp 1435–1440

Lele PP, Goddard J (1987) Optimizing insonication parameters in therapy planning for deep heating by SIMFU. Proceedings of the 9th annual conference of IEEE Engineering in Medicine and Biology Society, Boston, Nov 1987. IEEE, New York, pp 1650–1651

Lele PP, Parker KJ (1982) Temperature distributions in tissues during local hyperthermia by stationary or steered beams of unfocussed or focussed ultrasound. Br J Cancer 45 (Suppl V):108–121

Li GC, Hahn GM, Tolmach LJ (1977) Cellular interaction by ultrasound. Nature 267:163–165

Lizzi FL, Coleman DJ, Driller J, Ostromogilsky M, Chang S, Greenall P (1984) Ultrasonic hyperthermia for ophthalmic therapy. IEEE Trans Sonics Ultrason SU 31:473–481

Lyons M, Parker KJ (1988) Absorption and Attenuation in Soft Tissues II – Experimental Results. IEEE Trans Ultrason Ferroldectr Freq Control 35:511–521

Madsen EL, Zagzebski JA, Banjavie RA, Jutila RE (1978) Tissue mimicking materials for ultrasound phantoms. Med Phys 5:391–394

Madsen EL, Goodsitt MM, Zagzebski JA (1981) Continuous wave generated by focussed radiators. J Acoust Soc Am 70:1508–1517

Madsen EL, Zagzebski JA, Frank GR (1982) Oil-in-gelatine dispersion for use as ultrasonically tissue mimicking materials. Ultrasound Med Biol 8:277–287

Madsen EL, Sathoff HJ, Zagzebski JA (1983) Ultrasonic shear wave properties of soft tissues and tissuelike materials. J Acoust Soc Am 74:1346–1355

Marchal C, Bey P, Metz R, Gaulard ML, Robert J (1982) Treatment of superficial human cancerous nodules by local ultrasound hyperthermia. Br J Cancer 45 (Suppl V):243–245

Marmor JB, Nagar C, Hahn GM (1977) Tumor regression and immune recognition after localized ultrasound heating. Radiat Res 70:633

Marmor JB, Pounds D, Hahn N, Hahn GM (1978) Treating spontaneous tumors in dogs and cats by ultrasound-induced hyperthermia. Int J Radiat Oncol Biol Phys 4:967–973

Marmor JB, Pounds D, Postic TB, Hahn GM (1979) Treatment of superficial human neoplasms by local hyperthermia induced by ultrasound. Cancer 43:188–197

Martin CJ, Law ANR (1983) Design of thermistor probes for measurement of ultrasound intensity distributions. Ultrasonics 21:85–90

Martin CJ, Pratt BM, Watmough DJ (1983) Observations of ultrasound-induced effects in the fish Xiphophorous macalatus. Ultrasound Med Biol 9:177–183

Martin CJ, Hynynen K, Watmough DJ (1984) Measurement of ultrasound energy density distributions in vivo. Ultrasound Med Biol 10:701–708

Mason WP (1950) Piezoelectric crystals and their application to ultrasonics. Van Nostrand, Princeton

Mayer WG (1965) Energy partition of ultrasonic waves at flat boundaries. Ultrasonics 3:62–68

Moros EG, Roemer RB, Hynynen K (1988) Simulations of scanned focussed ultrasound hyperthermia: the effect of scanning speed and pattern. IEEE Trans Ultrason Ferroelectr Freq Control 35:552–560

Moros EG, Roemer RB, Hynynen K (1989) Pre-focal plane high temperature regions induced by scanning focussed ultrasound beams. Int J Hyperthermia (in print)

Mortimer AJ (1982) Physical characteristics of ultrasound. In: Repacholi MH, Benwell DA (eds) Essentials of medical ultrasound. Humana, Clifton

Munro P, Hill RP, Hunt JW (1982) The development of improved ultrasound heaters suitable for superficial tissue heating. Med Phys 9:888–897

Nakahara W, Kabayashi R (1934) Biological effects of short exposure to supersonic waves: local effects on the skin. Jpn J Exp Med 12:137

NCRP (1983) Biological effects of ultrasound: mechanisms and clinical implications. Report no 74. National Council on Radiation Protection and Measurements, Bethesda, MD

Nelson PA, Herrick JF, Krusen FH (1950) Temperatures produced in bone marrow, bone and adjacent tissues by ultrasound diathermy. Arch Phys Med 31:687–695

Nyborg WL (1981) Heat generation by ultrasound in a relaxing medium. J Acoust Soc Am 70:310–312

Nyborg WL, Ziskin MC (eds) (1985) Biological effects of ultrasound. Churchill Livingstone, New York (Clinics in diagnostic ultrasound, vol 16)

Ocheltree KB, Benkeser JP, Frizzell LA, Cain CA (1984) An ultrasound phased array applicator for hyperthermia. IEEE Trans Sonics Ultrason SU 31:526–531

Oka M (1960) Surgical application of high-intensity focused ultrasound. Clin All Round (Jpn) 13:1514

O'Neil HT (1949) Theory of focussing radiators. J Acoust Soc Am 21:516–526

Parker KJ (1983) The thermal pulse decay technique for measuring ultrasonic absorption coefficients. J Acoust Soc Am 74:1356–1361

Pennes HH (1948) Analysis of tissue and arterial blood temperatures in the resting human forearm. J Appl Phys 1:93–122

Ristic VM (1983) Principles of acoustic devices. Wiley, New York

Robinson TC, Lele PP (1972) An analysis of lesion development in the brain and in plastics by high-intensity focussed ultrasound at low-megahertz frequencies. J Acoust Soc Am 51:1333–1351

Roemer RB, Hynynen K, Johnson C, Kress R (1986) Feedback control and optimization of hyperthermia heating patterns: present status and future needs. Proceedings of the 8th annual IEEE/EMBS meeting, pp 1496–1499

Roemer RB, Swindell W, Clegg ST, Kress RL (1984) Simulation of focussed, scanned ultrasound heating of deep-seated tumors: the effect of blood perfusion. IEEE Trans Sonics Ultrason SU 31:457–466

Sehgal CM, Bahn RC, Greenleaf JF (1984) Measurement of the acoustic nonlinearity parameter B/A in human tissues by thermodynamic method. J Acoust Soc 76:1023–1029

Sehgal CM, Brown GM, Bohn RC, Greenleaf JF (1986) Measurement and use of acoustic nonlinearity and sound speed to estimate composition of excised livers. Ultrasound Med Biol 12:865–874

Seppi E, Shapiro E, Zitelli L, Henderson S, Wehlau A, Wu G, Dittmer C (1985) A large aperture ultrasonic array system for hyperthermia treatment of deep-seated tumors. Proceedings of the IEEE Ultrasonic Symp, pp 942–949

Shimm DS, Hynynen K, Anhalt DP, Roemer RB, Cassady JR (1988) Scanned focussed ultrasound hyperthermia: preliminary clinical results. Int J Radiat Oncol Biol Phys 15:1203–1208

Sommer FG, Pounds D (1982) Transient cavitation in tissues during ultrasonically induced hyperthermia. Med Phys 9:1–3

Stewart HF (1982) Ultrasonic measurement techniques and equipment output levels. In: Repacholi MH, Benwell DA (eds) Essentials of medical ultrasound. Humana, Clifton, NJ, pp 77–116

Stockdale HR, Hill CR (1976) Use of sphere radiometer to measure ultrasonic beam power. Ultrasound Med Biol 2:219–220

Swindell W (1985) A theoretical study of nonlinear effects with focussed ultrasound in tissues: an acoustic Bragg peak. Ultrasound Med Biol 11:121–130

Swindell W (1986) Ultrasonic hyperthermia. In: Hand JW, James JR (eds) Physical techniques in clinical hyperthermia. Research Studies, Letchworth, pp 288–325

Swindell W, Roemer RB, Clegg ST (1982) Temperature distributions caused by dynamic scanning of focussed ultrasound transducers. Proceedings of the IEEE Ultrasonic Symp (IEEE No 0090-5607), pp 745–749

Szent-Gorgyi A (1933) Chemical and biological effects of ultrasonic radiation. Nature 131:278

ter Haar GR, Stratford IJ, Hill CR (1980) Ultrasonic irradiation of mammalian cells in vitro at hyperthermic temperatures. Br J Radiol 53:784–789

ter Haar GR, Daniels S, Eastaugh KC, Hill CR (1982) Ultrasonically induced cavitation in vivo. Br J Cancer 45 (Suppl V):151–155

Tobias J, Hynynen K, Roemer R, Guthkelch AN, Fleisher AS, Shivley J (1987) An ultrasound window to perform scanned focussed ultrasound hyperthermia treatments of brain. Med Phys 14:228–234

Underwood HR, Burdette EC, Ocheltree KB, Magin RL (1987) A multielement ultrasonic hyperthermia applicator with independent element control. Int J Hyperthermia 3:257–267

Walker DCB, Lumb RF (1964) Piezoelectric probes for immersion ultrasonic testing. Appl Mater Res 3:176–183

Washington ABG (1961) Design of ultrasound probes. Br J Non Destr Test 3:56–63

Wells PNT (1969) Physical principles of ultrasonic diagnosis. Academic, London

Wells PNT (1977) Biomedical ultrasonics. Academic, London

Westermark F (1898) Über die Behandlung des ulcerireden Cervixcarcinomas mittels konstanter Wärme. Zentralbl Gynaekol 22:1335–1339

Williams AR (1983) Ultrasound: biological effects and potential hazards. Academic, London

Woeber K (1965) The effect of ultrasound in the treatment of cancer. In: Kelly E (ed) Ultrasonic energy: Biological investigations and medical applications. University of Illinois Press, Urbana, pp 135–149

Yoshioka K, Oka M (1965) Technical developments of focussed ultrasound and its biological and surgical applications in Japan. In: Kelly E (ed) Ultrasonic energy: biological investigations and medical applications. University of Illinois Press, Urbana, pp 190–201

Zemanek J (1971) Beam behavior within the nearfield of a vibrating piston. J Acoust Soc Am 49:181–191

3 Physics Evaluation and Quality Control of Hyperthermia Equipment

P.N. Shrivastava and T.K. Saylor

3.1 Common Components of Hyperthermia Equipment

A typical configuration of modern hyperthermia equipment used in clinics for localized cancer treatment is shown in Fig. 3.1. It can be described as a closed loop system in that the energy deposited into tissues is controlled by a feedback mechanism depending on the temperature of one or more reference probes. The applicator in this system is designed to deposit electromagnetic or ultrasonic energy into tissues which, when absorbed, is converted into heat and raises the temperature of the tumor. The overall goal is to elevate the temperature of the entire tumor to above 42 °C while maintaining the hot spots in normal tissues at below 46 °C for periods of up to 1 h. Two to ten treatment sessions over a period of 6 weeks may be required for complete therapy. Clinical experience to date has shown that even this seemingly modest goal is difficult to achieve

in a reliable and reproducible manner. The reasons for this are manifold. First, there is a lack of standard equipment to deposit energy in a controlled manner and generate optimum heating patterns in specific tumor sites. Second, the tissue-heating process depends on many dynamic physical and physiological mechanisms which are difficult to measure and control. Although factors like the form of energy, i.e., electromagnetic or ultrasound, the frequency of radiation, and the applicator design can be preselected to optimize heating patterns, other factors like the efficiency of power coupling and heat losses in tissues can change from treatment to treatment and alter the heat patterns achieved. The situation is further complicated if the temperature readings in the reference probes used to control power delivery are inaccurate or unreliable. The factors which affect reliability of temperature include design limitations, perturbations due to interaction of the probe with the energy fields, and artifacts like self-heating and conductive heat losses. Because of the many uncon-

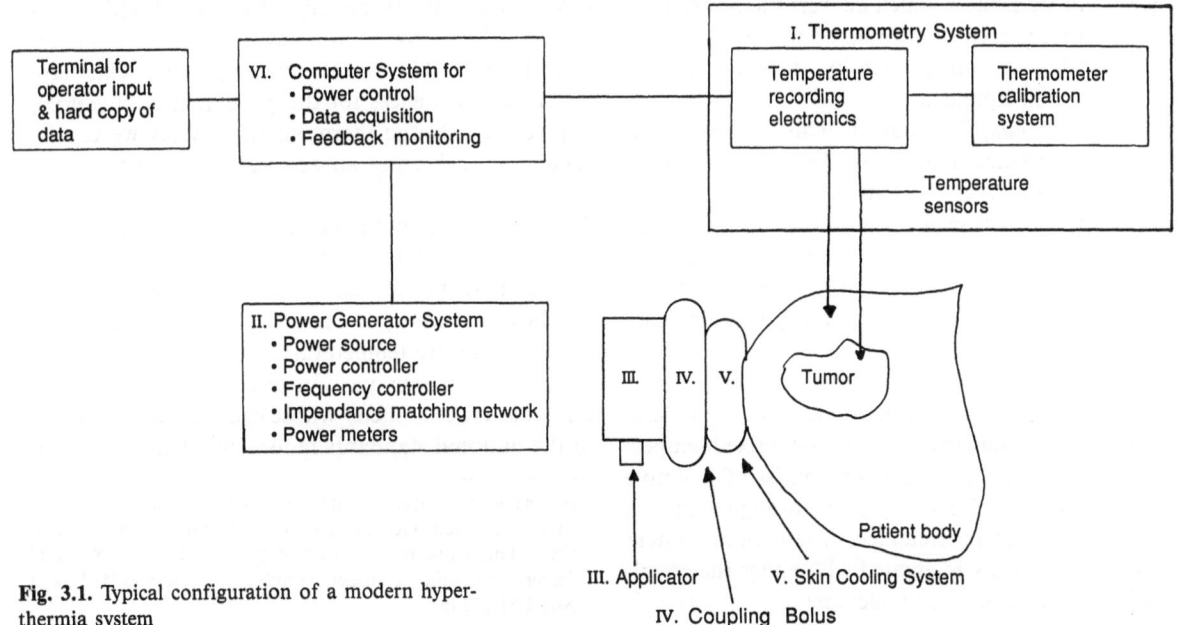

Fig. 3.1. Typical configuration of a modern hyperthermia system

trollable factors that can result in erroneous treatment delivery, it is essential to use stringent quality control measures so that each and every equipment-related factor and the predictable artifacts are controlled within strict specifications. This chapter describes the quality control criteria and procedures for each of the components of a hyperthermia system. The quality assurance guidelines for treatment delivery and documentation (Shrivastava et al. 1989) must be followed in addition to achieve safe and effective hyperthermia treatments. The six major components of the hyperthermia equipment in Fig. 3.1 are as follows:

1. *The thermometry system:* This consists of the temperature sensors, the leads which carry signals, the electronics for temperature recording, a calibration mechanism, and some computer-controlled software to provide feedback to the power controller and to keep temperature-time records for a number of temperature probes.
2. *The power generator system:* This consists of a radiofrequency or microwave power generator whose power output and frequency can be controlled over some fixed range either manually or by an automated mechanism driven by a feedback signal from the thermometry system. It includes the impedance matching network which transfers the power from the generator to the appropriate applicator, antenna, electrode, or transducer, which is coupled to the patient's body. The power meters and indicators are also a part of the generator system.
3. *The applicator:* This consists of a waveguide, antenna, or transducer which delivers the power over some surface area or volume of tissue. The exact distribution of power density depends on the design of the applicator.
4. *The power coupling system:* This consists of a suitable interface material (called bolus) between the applicator and the patient's body surface to facilitate power transfer from the applicator to the body. The efficiency of power transfer depends on the bolus material, size, thickness, etc. A high forward power and a low reflected power indicate good coupling efficiency.
5. *The cooling system:* With some equipment or in some applications, it is desirable to cool the skin surface or the underlying fat to some fixed temperature. The cooling system can consist of a simple air blower or of cooling pads through which a precooled fluid is circulated. The cooling system may be physically integrated with either the coupling bolus or even the applicator.

6. *A computer system for power control and data acquisition:* This consists of a computer and associated software which accepts operator input from a keyboard, monitors the thermometer readings and feedback signals, controls the power according to some algorithm, records the temperature-time data onto a disc, and displays them on a video screen or provids a hard copy on paper.

In order to produce effective and safe hyperthermia treatments, it is essential that all the components of a system be thoroughly tested to ensure that they meet the minimum specifications for acceptable quality equipment. The tests necessary to evaluate hyperthermia equipment and the criteria of acceptability described in this chapter are based on the experience of the Hyperthermia Physics Center[1] (HPC), which is responsible for developing a national Hyperthermia Quality Assurance Program in the United States.

3.2 Thermometry Evaluations

3.2.1 The Importance of Temperature Accuracy

The accuracy and reliability of the temperatures achieved during treatment are certainly the most important factors in ensuring the effectiveness of hyperthermia and patient safety. Based on the Arrhenius plots and the dose-response concept proposed by Dewey et al. (1977), heating of tissues at 42.7 °C instead of 42.5 °C for 1 h will have a 1.2 times higher effect. Thus, a 0.2 °C error in temperature can have serious consequences for the patient. It is, therefore, imperative that the thermometry devices we use be capable of achieving an accuracy much better than 0.2 °C.

The types of thermometry devices commercially available for clinical temperature monitoring and their characteristic performances (Shrivastava et al. 1988 a) are shown in Table 3.1. The accuracy with which these systems measure temperature can be defined on the basis of an intercomparison of their readings against a standard thermometer whose calibration is traceable to the national standards laboratory. The differences

---
[1] Hyperthermia Physics Center, Allegheny Singer Research Institute, Allegheny General Hospital, Pittsburgh, Pennsylvania, USA. This work is supported in part by the U.S. National Cancer Institute contract numbers N01-CM-37512 and N01-CM-87245.

Table 3.1. Measured characteristics of thermometers available commercially

	Thermocouples[a]	Bowman[b]	GaAs[c]	Fluoroptic[d]
Accuracy over 30° – 50 °C				
Mean error	+ 0.09	− 0.01	+ 0.15	− 0.27
SD of errors	0.32	0.33	0.32	0.30
Range (low, high)	− 1.0, + 3.9	− 4.6, + 3.1	− 2.7, + 1.0	− 0.9, + 0.4
Stability over 1 – 3 h period (°C/h)				
Mean drift/h	+ 0.02	+ 0.01	+ 0.16	+ 0.01
Root mean square	0.03	0.02	0.30	0.12
Range	− 0.04, + 0.05	− 0.03, + 0.07	− 0.99, + 0.54	− 0.09, + 0.23
Precision of 10 successive readings (°C)				
Mean of SD	0.019	0.013	0.036	0.050
Range of SD	0, 0.10	0, 0.08	0, 0.07	0.01, 0.11
Time (s) required to record 95% of temperature ("response time")				
Bare probes				
Mean	4.5	2.4	5.7	2.1
Range	4.4, 4.6	0.65, 3.5	4.6, 6.6	1.6, 3.0
In catheters				
Mean	6.4	5.9	10.3	–
Range	5.6, 8.8	1.7, 8.1	5.5, 24.4	–

[a] Thermocouples from various manufacturers
[b] BSD Medical, Inc., Salt Lake City, Utah
[c] Clini-Therm Corp., Dallas, Texas
[d] Luxtron Inc., Mountain View, California. Model 2000 is now replaced by the improved model 3000

in readings T_i of the test and T_s of the standard thermometers under identical conditions measure the lack of accuracy.

The frequency distribution of 2458 intercomparisons made by the HPC for 166 thermometers, including thermocouples, thermistors, Bowman probes, GaAs probes, and fluoroptic probes, is shown in Fig. 3.2.

These intercomparisons were made in the 29.772° to 50 °C temperature range. It is seen that the mean temperature error is centered near 0 °C with a standard deviation of 0.34 °C. Note, however, that this distribution is not strictly Gaussian and that 74% of the data points fell within − 0.25° to + 0.25 °C even though the standard deviation was 0.34 °C. The remaining 26%

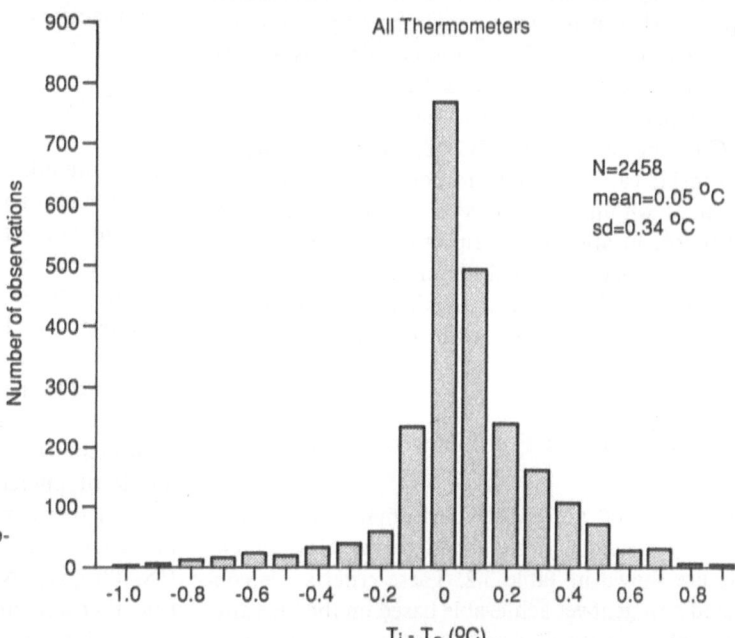

Fig. 3.2. Distribution of temperature errors observed in 2458 intercomparisons for 166 thermometers including thermocouples, thermistors, Bowman probes, GaAs probes, and fluoroptic probes

All Thermometers

N=2458
mean=0.05 °C
sd=0.34 °C

$T_i - T_s$ (°C)

Number of observations

of the data points fell in a broad range extending from $-4.6°$ to $+3.9°C$ (extremes not shown in Fig. 3.2). These results clearly indicate that the goal of $0.2°C$ accuracy in the clinic is unlikely to be achieved without stringent quality control measures. In our survey we also found that the unacceptable inaccuracies above $0.2°C$ were often caused by faulty calibration procedures, insufficiently frequent calibrations, inherent short- or long-term drifts, artifacts, or some systemic limitation of the thermometry system tested. Thus, the quality assurance (QA) procedures at each facility must include not only a thorough laboratory evaluation of the thermometry system to assess accuracy achievable, and any physical limitations, but also adherence to the correct procedures necessary for preventing or minimizing systemic errors and other artifacts during treatment. The laboratory tests and the guidelines for proper use of the most commonly available thermometers are described below.

3.2.2 Scope of Laboratory Tests

Laboratory tests of a thermometry system should check the system accuracy, precision, stability, response time, probe diameter, and sensor location within the probe. As already mentioned, "accuracy" is determined by comparison with a standard calibrated thermometer. "Precision" is determined by the standard deviation of ten successive readings of a thermometer in a stable environment. "Stability" is characterized by the drift per hour of a thermometer response when it is used in environments simulating the patient for $1-3$ h. The "response time" is defined by the time required to register 95% of the temperature change when a probe is suddenly moved from $25°C$ to approximately $45°C$. If probes are clinically used inside catheters, their response times should be measured within the catheters. The probe diameter and sensor location with respect to its tip are simple measurements of practical importance to ensure that the probe will fit within the catheter and to specify sensor position accurately with respect to anatomy.

3.2.3 Criteria and Frequency of Testing

The criteria of acceptable performance of a thermometry system and suggested frequencies for QA tests are shown in Table 3.2. These criteria are considered stringent yet achievable based on the measurements of a variety of equipment at a number of in-

stitutions reviewed by the Hyperthermia Quality Assurance Program in the United States.

3.2.4 Testing Procedures

All thermometers should be compared to a standard thermometer in a precisely controlled water bath. Each institution should have a calibrated secondary standard thermometer as a reference. Precision mercury-in-glass thermometers readable to $0.1°C$ in the $20°-50°C$ range and having calibrations traceable to a national standards laboratory are generally adequate and quite inexpensive. A drawback of these thermometers is that because of their size they often cannot be used in small cavities like those provided by manufacturers for calibration of clinical probes. It is, therefore, convenient to have a local tertiary standard thermometer consisting of a thermocouple or thermistor. Our tertiary standard thermometer consists of a thermistor (Thermometrics, Inc., model A128-CSP60BT103M) with a digital multimeter (Hewlett Packard 3478A) used to a sensitivity of 1 ohm. The tertiary standard is closely intercompared to the secondary standard and is estimated to have worst case inaccuracy within $±0.03°C$. T_i and T_s in the discussions below refer to readings of the test and standard thermometers under identical situations. The difference $T_i - T_s$ is a measure of thermometer inaccuracy. If the inaccuracy exceeds the indicated criteria for any test, corrective actions should be taken. Corrective actions usually call for recalibration or replacement of the clinical thermometers.

3.2.4.1 Accuracy of Calibration Thermometers

Some manufacturers provide calibration thermometers with their thermometry system against which the clinical thermometers are calibrated. However, we recommend that each institution have its own standard thermometer, which will serve as an independent reference for both clinical and calibration thermometers. All calibration thermometers should be compared to the reference standard at least at ten different temperatures in the $20°-50°C$ range. The agreement should be within $0.1°C$. Table 3.3 is an example of intercomparisons made for two BSD calibration thermistors against the HPC standard thermometer in an ultrahigh precision water bath (Exacal EX-100 UHP, Neslab, Fortunate, NH). Both calibration thermometers met the accuracy criterion at all data points measured.

Table 3.2. Recommended criteria and frequency for thermometry QA tests

Test/procedure	Acceptability criteria	Frequency[a]
A. Thermometry		
1. Calibration thermometers:		
Accuracy over $20° - 50°C$	$0.1°C$	A, Y
2. Clinical thermometry:		
Accuracy over $30° - 50°C$	$0.2°C$	A, Q
Precision	$0.1°C$	A, Q
Stability	$0.1°C/h$	A, Q
Response time	$<10\,s$	A, Y
Accuracy of sensor position	$2\,mm$	A, Q
Perturbations/artifacts	$< 0.1°C$	A, C, Y
EM interference	$< 0.1°C$	A, C, Y
Pretreatment accuracy check at $\leq 37°C$ and $\geq 43°C$	$0.1°C$	T or R
Inspection of probe damage	Qualitative	A, T

[a] The frequency codes are:
    A = Acceptance testing and equipment installation
    Y = Yearly
    Q = Quarterly
    T = Each treatment
    R = Recommended by manufacturer
    C = When some components of a standard setup are changed

Table 3.3. Calibration thermometer accuracy[a]

T_s (°C)	BSD thermistors $(T_i - T_s)$ (°C)	
	A	B
28.615	$+0.03$	$+0.03$
30.979	$+0.02$	$+0.05$
34.343	$+0.02$	$+0.06$
36.728	$+0.04$	$+0.02$
39.619	$+0.01$	$+0.02$
42.241	0.0	$+0.01$
45.089	-0.02	$+0.03$
47.387	0.0	0.0
50.542	$+0.04$	-0.02
29.771[b]	$-$	$+0.08$

[a] The HPC recommends that institution calibration thermometers agree with T_s to within $\pm 0.1°C$ over the range $20° - 50°C$. This criterion is met by both thermometers at all data points measured
[b] In gallium cell

3.2.4.2 Accuracy of Clinical Thermometers

All thermometers used in patients should first be properly calibrated as specified by the manufacturers and then compared with the standard thermometer at least at ten temperatures in the 30°–50°C range. The agreement should be within 0.2 °C. Some clinical thermometers are sensitive to moisture and should be placed within closed-end catheters before insertion in water baths. Even those thermometers which are not sensitive to moisture should be tested within catheters if they are normally used within catheters in patients. An example of this test for an eight-channel system (BSD-1000) is shown in Table 3.4. It should be noted that probes 3 and 4 did not meet the 0.2 °C accuracy criterion while probes 7 and 8 failed during the test. It is important to recognize that each probe in a multiprobe system has to be tested individually and that interchanging a probe and its readout channel can result in miscalibration.

3.2.4.3 Precision of Clinical Thermometers

Ten successive readings, approximately 10 s apart in a stable water bath near 42.5 °C, should be recorded for each clinical probe. The calculated values of the mean, the standard deviation, and the actual low and high readings as shown in Table 3.5 indicate the precision achievable with each probe. We recommend that the standard deviations for clinical probes should not exceed 0.1 °C. This criterion was met by all probes in the test.

3.2.4.4 Stability of Clinical Thermometers

To measure stability the initial readings T_i for each clinical probe should be taken at some fixed tempera-

Table 3.4. Accuracy of clinical thermometers: Bowman probes − BSD-1000

T_s (°C)	$(T_i - T_s)$ (°C)							
	1	2	3	4	5	6	7	8
28.615	+0.08	+0.08	−0.21	−0.31	+0.08	+0.08	+0.08	+0.08
30.979	+0.02	+0.02	−0.18	−0.48	+0.02	+0.02	+0.02	+0.02
34.343	−0.04	−0.04	−0.34	−0.64	−0.04	+0.06	−0.04	−0.04
36.728	−0.03	−0.03	−0.13	−0.33	−0.03	−0.03	−	−
39.619	−0.02	−0.02	−0.22	−0.72	−0.02	−0.02	−	−
42.241	+0.06	−0.04	−0.54	−0.64	−0.04	−0.04	−	−
45.089	+0.01	+0.01	−0.69	−0.89	+0.01	−0.09	−	−
47.387	+0.01	+0.01	−1.29	−1.39	+0.01	+0.01	−	−
50.542	+0.06	−0.04	−2.64	−2.64	−0.04	−0.04	−	−
29.771[a]	+0.03	+0.03	−0.67	−0.571	+0.03	+0.03	−	−

[a] In gallium cell

−, Thermometers 7 and 8 failed during the test

Table 3.5. Precision of clinical thermometers: precision at $T_s = 43.387$ °C; BSD Bowman probes

	A	B	1	2	3	4	5	6
Mean	47.375	47.390	47.4	47.4	46.13	45.91	47.4	47.32
SD	0.024	0.023	0.0	0.0	0.067	0.031	0.0	0.042
High	47.39	47.44	47.4	47.4	46.2	46.0	47.4	47.40
Low	47.34	47.34	47.4	47.4	46.0	45.9	47.4	47.3

Table 3.6. Stability of clinical thermometers (°C)

Probe	T_i	$(T_i - T_s)$	T_f	$(T_f - T_s)$	$(T_f - T_i)$
1	29.8	+0.03	29.8	+0.03	0.0
2	29.8	+0.03	29.8	+0.03	0.0
3	29.3	−0.47	29.1	−0.67	−0.20
4	29.2	−0.57	29.2	−0.57	0.0
5	29.8	+0.03	29.8	+0.03	0.0
6	29.8	+0.03	29.8	+0.03	0.0
7	29.8	+0.03	−		−
8	29.8	+0.03	−		−

−, Probes 7 and 8 failed during calibration procedure

ture near 42 °C in the water bath or in a gallium melting point cell (29.77 °C) if one is available. These measurements should be repeated after an interval of 1−2 h or longer if preferred to get final readings T_f. The difference $T_f - T_i$ indicates calibration drift during the interval that the probes are in clinical use. In situations where the standard waterbath temperature T_s does not remain steady during the interval between initial and final readings, it is best to intercompare $(T_i - T_s)$ and $(T_f - T_s)$ values to determine drift. For acceptable performance the drift per hour should be less than 0.1 °C. The data in Table 3.6 indicate at least one probe (# 3) with unacceptable drift. It may be noted that it is the same probe which failed the accuracy test in Table 3.4.

It is reported that the stability of some thermometers is affected by tissue moisture even when they are within closed-end catheters (Saylor et al. 1984, 1988). In order to ascertain that probe stability is not affected by moisture or prolonged interstitial use during patient treatments, it is prudent to maintain the thermometers either in a water bath or in tissues during the interval between T_i and T_f.

3.2.4.5 Response Time of Clinical Thermometers

"Response time" is the time required to register 95% of the temperature change when a probe is suddenly moved from approximately 25 °C to approximately 42.5 °C. If the probe is clinically used in a catheter, its response time should also be measured in a catheter. A typical time-response curve for a photoluminescent probe is shown in Fig. 3.3. It is seen that the response time is less than 1 s. However, for some probes that we have tested, although the response time may be much less than 1 s, the time to record or print the temperature output can exceed 10 s. It is the overall reading time that is of importance. Response time of clinically acceptable probes should be less than 10 s.

3.2.4.6 Probe Diameter and Sensor Position

The probe diameter and sensor position involve simple mechanical measurements to be made with a calliper or ruler. In clinical work it is important that the invasive temperature probes be of small diameter so as to minimize tissue trauma. Among the commercially available probes used clinically to date, the Bowman probes have the largest diameter (1.1 mm) and are insertable in an 18-gauge catheter. Since these probes can satisfy all the other criteria of acceptable

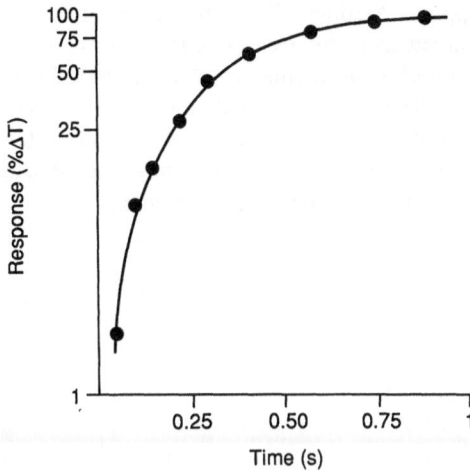

Fig. 3.3. A typical time-response curve for a fluoroptic thermometer. Δ T is the temperature change to which the thermometer is subjected

performance, it is recommended that any new probes have diameters less than 1.1 mm.

The sensor position with respect to the probe tip is a necessary piece of information to specify accurately the point at which the temperature is measured. We have observed inconsistent displacements of as much as 0.5 cm in a set of probes available with a thermometry system. It is well recognized that half centimeter inaccuracies in probe position, especially near the periphery of a tumor, can give erroneous clinical information. It is, therefore, recommended that the sensor location be known to an accuracy better than 2 mm and be periodically checked to ensure that no displacements have occurred.

3.2.4.7 Perturbations and Artifacts

It is well recognized that temperature probes interfere and interact with the microwave (MW), radiofrequency (RF), and ultrasound (US) fields in a complex way and can potentially produce erroneous readings. These errors, classified as perturbations when present, are superimposed on top of the inaccuracies measured under ideal water bath conditions described above.

Perturbations can be qualitatively considered to result from two causes: (a) change in the probe readings due to presence of the field and (b) change in the field due to presence of the probe. In practice the effects of these two components are difficult to separate. At least some perturbation effects have been observed for each type of clinically used thermometer under some irradiation conditions.

In addition to perturbations, artifacts are another potential source of temperature inaccuracies. Artifacts are defined as systematic temperature errors due to extraneous, environmental, or some other stressful condition affecting probe readings. With unfiltered thermocouples in MW fields, these effects are overwhelming and obvious but with other nonperturbing systems they may remain unrecognized even though present. A number of laboratory tests are possible to check whether a particular type of thermometer is associated with perturbation or artifactual inaccuracies. However, these tests are often complex and require specialized equipment so as to be impractical for routine QA tests in most clinics. Also, the extent of inaccuracies caused by the perturbation or artifact phenomena depends on specific local conditions like the magnitude of the power applied, the method of power coupling, and the orientation of the electric field (E-field) with respect to the thermometer probe. Hence, these errors may be present in one patient setup and not in another. In practical clinical applications, therefore, it is preferable to recognize the conditions under which perturbation effects occur with a given thermometer and then use a preventive QA strategy to minimize the possibility of them becoming a large source of error. The phenomena of perturbation and artifacts and their effect on temperature accuracy are discussed in detail below.

3.2.5 Some Known Sources of Temperature Errors

3.2.5.1 Electromagnetically Induced Currents in Metallic Probes

The electromagnetic (EM) interference in metallic probes (Christensen 1983) is generally caused by the currents induced in the conductive metals and/or leads by the incident field. These EM-induced alternating currents cause three types of difficulty: (a) they, themselves, become a new source of radiation, heating the surrounding tissues; (b) they resistively heat the probe; and (c) they reach the measurement instrument via leads and interfere with the signal processing circuits, causing erratic readings. The magnitude of the induced currents depends on the magnitude of the incident E-field parallel and perpendicular to the interface between tissues and the metallic elements. In the case of thin metal wires, the parallel component is the most significant problem. Therefore, the magnitude of this effect can be reduced somewhat by carefully aligning the long axis of the probe perpendicular to the incident E-field. This makes the induced currents short in length, reduces self-heating, and also minimizes reirradiation effects. Unfortunately, in practice it is not always possible to know the orientation of the incident E-field either because such orientation does not exist, as in the near-field region of many applicators, or because refractions and reflections by intermediate tissues disrupt the polarization to an unknown direction. These problems are most prominent when thermocouples are used in EM fields.

In spite of these perturbation effects, thermocouples are used by some investigators for monitoring EM-induced hyperthermia. Some investigators place probes perpendicular to the E-field during heating, and most turn off the incident field to read temperatures without EM interference of the instruments. To avoid inaccurate readings resulting from self-heating of the probe, it is necessary to wait a few seconds after the power is turned off and before temperature readings are taken. The waiting interval, which depends on the probe construction and the extent of self-heating, must be experimentally determined for each individual situation.

3.2.5.2 Viscous Heating and Ultrasonic Absorption

Thermocouples are largely free of electronic perturbations in ultrasonic (US) fields and to date are the only practical option for thermometry in patients heated with US devices. However, in this case two other artifacts, i.e., viscous heating and local US absorption, can affect their accuracy.

Viscous heating results from shearing forces caused by streaming of fluids over the probe surface when probes are tested or calibrated in liquid environments (Curley and Lele 1984). High local US absorption in plastics used for coatings on thermocouples or in the construction of catheters is another potential source of artifactual temperature errors (Fessenden et al. 1984). Unless preventive measures are taken, these effects can easily cause errors exceeding the desirable accuracy of 0.2°C.

3.2.5.3 Linear Mapping with Bare Thermocouples

Thermocouples are often inserted into hypodermic needles for tissue implantation and moved in tissues either for correct placement or for linear temperature mapping. Artifactually high temperatures have been observed with this technique (Fessenden et al. 1984). Presumably this artifact is related to blood clotting at the probe end or to severe damage to local tissues near the probe tip. This effect can be minimized by avoiding sharp, ragged tips and utilizing a suitable catheter whenever linear mapping is required.

3.2.5.4 Temperature Smearing Across High Gradients

If a probe is metallic and hence a much better conductor of heat than surrounding tissues, it can distribute the heat along its length, causing artifactual smearing of temperature readings when it is used in regions of high temperature gradients. Such an effect has been reported (Fessenden et al. 1984). This effect can be minimized by utilizing very thin metallic probes or prevented by utilizing nonconducting probes whenever possible.

3.2.5.5 Cross-talk in Multijunction Probes

Multiple junction thermocouples are now commercially available to monitor temperatures along a linear track without requiring probe motion. Cross-talk and heat gradient effects are possible sources of error with such probes. Errors of up to 1°C were reported for a ten-junction probe, due to cumulative thermoelectric potentials caused by nonisothermal conditions over thermocouple solder junctions in high thermal gradients (Lee and Fessenden 1984). In a three-junction probe mounted inside a 23-gauge hypodermic needle, the error was only 0.25°C in a temperature gradient as high as 3.7°C/mm (Curley and Lele 1984). In clinical applications it is necessary to be aware of these problems and to avoid the use of multijunction thermocouples unless their reliability has been verified.

3.2.5.6 Calibration Inaccuracies

Almost all clinical facilities check the reliability of their thermometry systems in one way or another before starting patient treatments. However, for various reasons, these tests are often unreliable without arousing suspicion of their inadequacy. A major source of unreliability stems from the fact that the probe calibrations are often checked against the same standard by which they are calibrated. A second source of uncertainty derives from the practice of

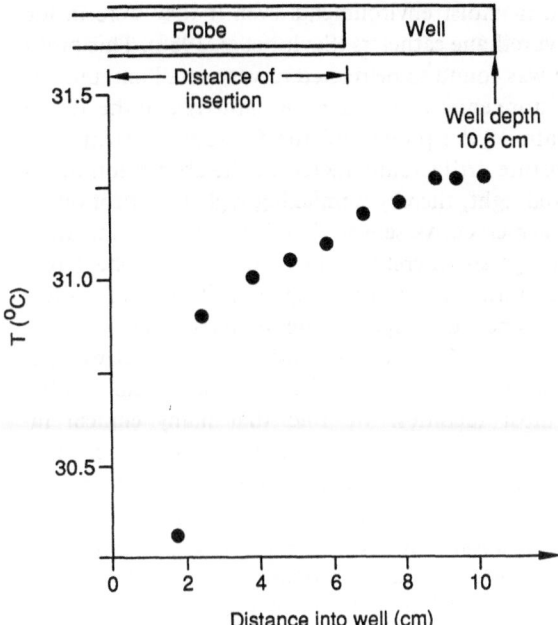

Fig. 3.4. Example of temperature variation inside a calibration well as a function of depth

often provide the clue as to the source of temperature miscalibrations and the period over which they occurred.

Certain multiprobe thermometry systems provide an option of interchanging probes and readout channels with or without requiring recalibration. Such interchangeability should not be accepted on faith but should be tested by placing all probes at some fixed temperature in a water bath or gallium melting point cell and verifying that interchanging channels does not change temperature indications.

3.2.5.7 Extrapolation Errors

The practice of checking reliability of a probe at a single temperature (e.g., room temperature) before using it in patients can under some circumstances be unreliable. This is because most automated calibrations are valid for a limited temperature range and even when the laboratory tests in water baths demonstrate the probes to be reliable, the room temperature readings may be beyond their accuracy range and include inherent extrapolation errors.

In order to prevent extrapolation errors, we propose that the pretreatment checks include intercomparisons against a standard thermometer at least at two temperatures, one just below ($< 37\,°C$) and another just above ($> 45\,°C$) the range of temperatures to be monitored during clinical use.

An example of errant behavior of two probes (#6 and #3), found to be accurate near $23\,°C$ but very inaccurate in the clinical range, is shown in comparison to the other well behaved probes of an eight-probe system in Fig. 3.5.

checking the calibration only at one temperature (e.g., room temperature).

Computerized, automatic calibration procedures rely on the accuracy of preassigned temperatures of calibration wells to generate a table of calibration constants which are used in polynomials to derive probe readings. If the actual well temperature differs from its preassigned value or lacks uniformity as a function of depth inside the well, the calibration constants can be in error. Figure 3.4 is an example of temperature variation within a well and demonstrates the need not only to check well temperatures but also to ensure that probes to be calibrated are fully inserted. Very often the calibration wells are too narrow to insert standard thermometers and to check their temperature accuracy and uniformity independently. If this is the case, an independent check of the reliability of the calibration procedure is impossible. One is then forced to rely only on the water bath tests described earlier. More frequent water bath tests should be required under these circumstances. Since probe miscalibrations can result from problems either in the calibration well or in the probes themselves, when the well cannot be checked, one may be inclined to replace probes even though the well is the real problem. With thermometry systems capable of printing the tables of calibration constants, we find it advisable to maintain a historical record of these constants because they can

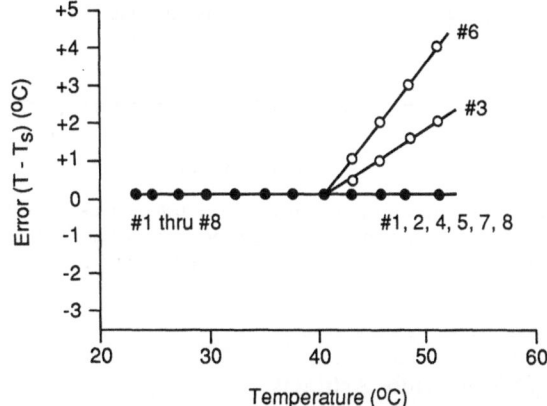

Fig. 3.5. Temperature errors observed for an eight-probe system over 22°–50°C. All eight probes in the system were accurate at room temperature (23°C) but two showed large errors above 40°C

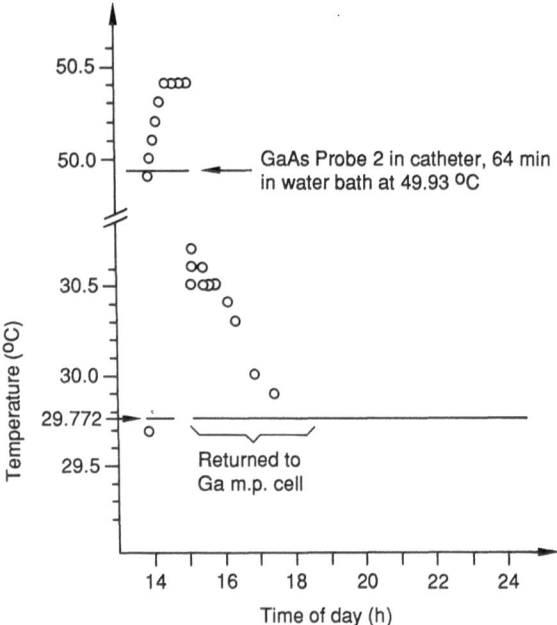

Fig. 3.6. Moisture artifact induced in a catheter-enclosed GaAs probe after 64 min in a water bath at 49.93 °C. The initially accurate probe shows artifactually higher temperature after use in a water bath. *Ga m.p. cell*, gallium melting point cell

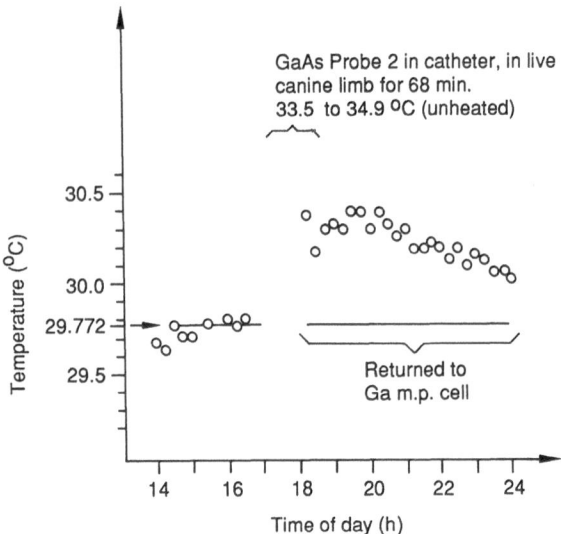

Fig. 3.7. Moisture artifact induced in a catheter-enclosed GaAs probe after 68 min interstitial insertion in a canine limb at 33.5°–34.9 °C. *Ga m.p. cell*, gallium melting point cell

3.2.5.8 Moisture Artifacts

In the case of the first generation GaAs fiberoptic probes (manufactured before 1986), a calibration drift of the order of 1 °C/h was observed when they were used in moist environments even if they were inside polyurethane catheters (Saylor et al. 1988). This problem was found to be related to very small amounts of moisture entering through the catheter and the Teflon sheath of the probe into the fiberoptic system. The moisture artifactually increased the absorption of infrared light, thereby mimicking higher absorption by a warmer GaAs sensor. The artifactual temperature readings are, therefore, always higher than the actual temperature, as shown in Fig. 3.6. The manufacturers have since developed some protection against this problem in their new generation probes, but we continue to observe this artifact at least in some probes at most facilities. We find that many clinical investigators, even though aware of this artifact, have assumed that it is not a problem when probes are used inside closed-end catheters in tissues. Our laboratory investigations, however, showed that moisture contamination remained a problem even when the probes were used inside catheters in live canine tissues, as demonstrated in Fig. 3.7.

3.2.5.9 Electromagnetic Interference in Electronics

Electromagnetic interference caused by stray EM or RF fields impinging either directly or carried via cables to the electronic components of otherwise nonmetallic, noninterfering probes is a potential source of unsuspected inaccuracy in temperature readings (Saylor et al. 1988). The extent of the error depends on the stray power density reaching the electronics and the adequacy of the RF shielding or filtering included in the electronics. Thus, changes in the spatial arrangements of the electronic components, cables, and the applicator or the power settings and coupling techniques can change the size of this interference artifact. An example of this effect is shown in Fig. 3.8. A simple pretreatment QA test is recommended to check the presence or absence of this artifact. The test requires placing all the probes at some fixed temperature, for example in a cup of warm water, close to the applicator but outside the irradiating field, while the applicator itself is set on the patient or a phantom to simulate treatment conditions. With this setup and utilizing a power setting near the maximum to be used clinically, the readings of each probe should be checked with the power "on" and "off." Under ideal conditions, since there is no change in probe temperatures, there should be no change in the probe readings. However, if the stray EM fields affect the electronic circuits, a change in readings can be observed. If a problem exists, either

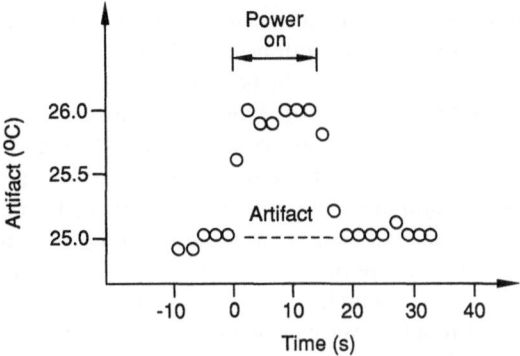

Fig. 3.8. The artifactual increase in temperature reading of a probe placed in a constant temperature water cup when the EM power is turned "on"

the cables or the electronic components can be repositioned or the RF shielding of the electronics can be improved to prevent or minimize this problem. In practice the inaccuracy due to EM interference should be less than 0.1 °C.

3.2.5.10 Probe Damage

Most clinical probes used in hyperthermia treatments are delicate and fragile. Physical damage of the tips has been observed to be associated with unacceptable inaccuracies in temperature readings during our field reviews of different institutions. A qualitative visual inspection of the probes before each treatment is, therefore, a prudent QA measure to reduce temperature inaccuracies. In some cases we have also seen that the sensor position was shifted with respect to the probe tip either due to differential thermal expansions or physical shielding. Such a shift can cause inaccuracies in defining the exact location of the sensor. The position of the sensor within a probe assembly should be periodically checked.

3.3 Power Evaluations

3.3.1 Electromagnetic Shielding and Electric Safety Requirements

Radiofrequency or MW power sources generating power in the range of 1 W to 1 kW are used in the heating systems available for hyperthermia treatments. In most countries certain regulations apply to the use of EM frequencies for medical purposes.

Generally the RF sources used to generate US power in transducers operating in the 0.3 – 3 MHz range are not regulated. In the United States the Federal Communications Commission (FCC) allows unrestrained use of EM power only at 13.52, 27.12, 915, and 2450 MHz for industrial, scientific, or medical (ISM) purposes. These frequencies are, therefore, referred to as the allowed ISM frequencies. In Europe in addition to the above, 433 MHz is also an allowed frequency. Because of these regulatory constraints, most hyperthermia devices are manufactured to operate at the ISM frequencies unless some technical advantages are foreseen in utilizing other frequencies. When other frequencies are used, FCC regulations (10 CFR 18.301) require that the power sources be registered as investigational devices, be enclosed in a metal screen-shielded room, and meet the recommendations regarding EM leakage.

Under certain circumstances the electric hazards of high power EM generators can be fatal to patients and operators. Therefore, the American National Standards Institute (ANSI) has established specific electric safety standards regarding ground faults, integrity of switches and connectors. These standards and the procedures to ensure that they are met are described in the ANSI publication *AAMI*, ES1, 1985. The recommended electric safety checks must be performed periodically. We suggest that these tests be performed by the clinical or biomedical engineers familiar with the test procedures and ANSI criteria rather than by the clinical staff.

In addition to these regulatory and safety tests, it is necessary to perform certain QA tests to ensure reliable, reproducible operation of the power sources and meters during patient treatments.

3.3.2 The Need for Accuracy of Power Readings

It is argued by some hyperthermia investigators that it is unimportant to know the power levels with any accuracy because during treatment the power is controlled by feedbacks from reference thermometers. However, since very often patient treatments are limited by the power available from the machine and since very high reflections can cause unsafe EM exposures to patient organs or to operators, we believe that prudent and accurate control of power delivered to patients is important and possible only if the power levels reaching the patient are accurately known.

In addition to improving safety, accurate knowledge of power used during each hyperthermia session can

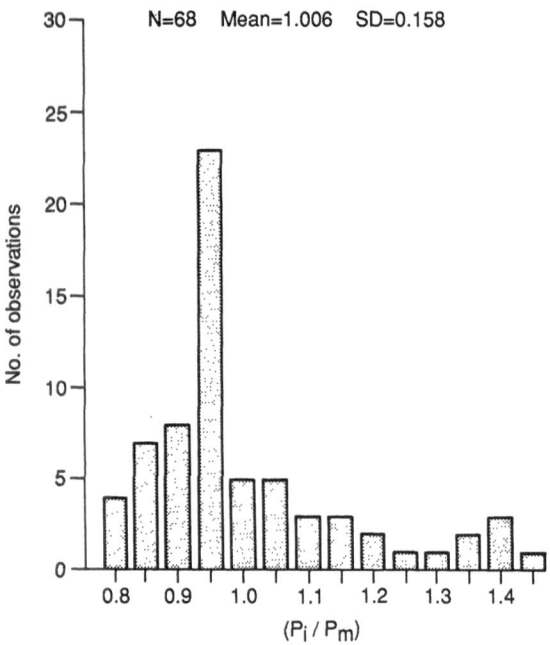

Fig. 3.9. Distribution of errors in indicated power (P_i) compared to measured power (P_m) for 68 intercomparisons on five generators operating between 13.56 and 1000 MHz

be very valuable in improving the reproducibility and reliability of fractionated patient treatments. For example, if one treatment session required unduly high power, it could indicate a divergence in the treatment setup and the need for prompt corrective actions. Furthermore, accurate record of the net power used in a treatment greatly improves the precise characteriza-

tion of the treatment technique and hence its transferability from one institution to another.

Accurate knowledge of the net power delivered into a medium by an applicator is also required in the measurements of specific absorption rate (SAR). Reliable measurements of SAR are needed to evaluate hyperthermia equipment, to intercompare their heating abilities, and to match applicators to heating requirements for specific tumors. The use of SAR data for treatment planning is based on the assumption that the input power can be precisely controlled.

Fig. 3.9 shows a histogram of some 68 measurements made by the HPC on five different generators spanning a 13.56 – 1000 MHz frequency range. The power indications P_i on the instrument panel, monitor, or computer are found to differ from those actually measured at the applicator by a large factor, ranging between 0.75 and 1.45. We find that these differences are generally due to a combined effect of zero offset on the panel meter, meter miscalibration, and inherent power losses in the line components between the generator and the applicator.

Actual measurements of power at the generator output P_G and at the applicator input P_A show that the panel meter indications P_i can usually be corrected by a formula:

$$P_G = A + B P_i + C P_i^2 \tag{3.1}$$

where the zero offsets, miscalibrations, and any nonlinearities in the panel meter are accounted for by the coefficients A, B, and C. The subsequent line

Table 3.7. Scope, criteria, and frequency of QA tests for power meters, applicators, and EM leakage

Test/procedure	Acceptability criteria	Frequency[a]
Power source and meter		
Electric safety	ANSI-AAMI ES1-1985	A, Y
EM and RF shielding	US CFR Part 18	A
Accuracy of net power into applicator	10%	A, C
Applicators and coupling devices		
Check phantom SAR data from manufacturer	10%	A, Y
Check heat patterns with coupling	Qualitative	A, C, Y
Single point SAR reproducibility	10%	T
EM hazard		
EM hazard meter calibration	Yearly certificate	A, R
EM exposure of operator	Within ANSI limits	A, C, Y
EM hazard surveys	Minimize unnecessary patient exposure	C or T

[a] The frequency codes are:

A = Acceptance testing and equipment installation
Y = Yearly
Q = Quarterly
T = Each treatment
R = Recommended by manufacturer
C = When some components of a standard setup are changed

losses between the generator and the applicator can also be accounted for by a transmission factor T such that

$$P_A = T \times P_G . \qquad (3.2)$$

From Fig. 3.9 we conclude that the accuracy of power readings in the present equipment is insufficient and suggest that all equipment manufacturers make every effort to provide accurate power indications on their instrument panels. In addition, every clinical facility must undertake certain QA measures to ensure that their records of net power at the applicator are accurate to within 10%.

3.3.3 Scope, Criteria, and Frequency of Power Meter Tests

In view of the importance of precise delivery and control of power to achieve reproducible, safe, and effective hyperthermia treatments, we recommend that each facility verify its power meter calibrations and line losses and account for them. A one-time determination of the constants A, B, C, and T and periodic verifications thereafter can ensure reliable accuracy, much better than 10%, in the net power delivered to the applicators during treatments.

A precisely calibrated, independent power meter should be used as a standard to measure power levels at both the generator output and the applicator input. The recommended tests, QA criteria, and frequency of testing are shown in Table 3.7. The testing procedures are described below.

3.3.3.1 Accuracy of Power Indicators

The standard power measuring system used by the HPC includes a power meter (HP 436A, Hewlett Packard, Inc., Palo Alto, CA), a power sensor (HP 8482A), two dual direction couplers (Amplifier Research DC 2000 and Narda 3020, Hauppauge, NY), and two attenuators (10 dB, HP 8491A and 20 dB, Weinschel 30-20-34). All the components have been precisely calibrated in the 50- to 2500-MHz frequency range so that the overall accuracy of power measurements is within 2%. These components are selectively used in a configuration shown in Fig. 3.10 to ensure that the sensor receives the correct power level within its sensitivity range.

This elaborate power measuring system is suitable for survey of a variety of equipment operating over a wide range of frequency and power outputs. However, at individual clinics such an elaborate and expensive system may be unnecessary and can usually be replaced with simple, relatively inexpensive power meters, calibrated either by the manufacturer or by a standard laboratory for a specific frequency over a limited power range to provide 2% – 5% measurement accuracy.

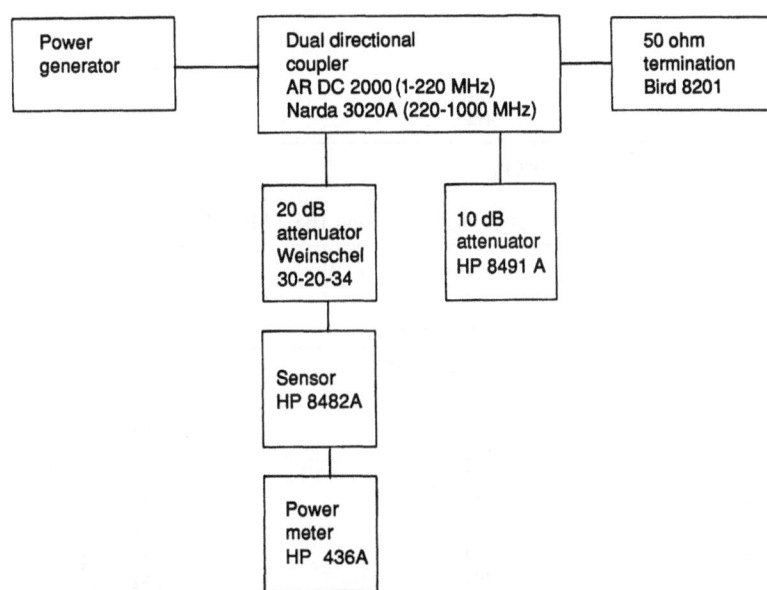

Fig. 3.10. Schematic configuration of the power measurement system used by the Hyperthermia Physics Center (HPC)

A standard calibrated power meter should be used to measure the power levels P_G at the generator output for a number of power settings P_i indicated on the console meter and covering the whole range of power produced by the equipment. Generally large deviations between the measured and indicated power values are observed, as shown in Table 3.8. However, these data can be used to derive the values of constants A, B, and C by a least square fit to the equation of the form

$$P_G = A + B P_i + C P_i^2 \ . \tag{3.3}$$

In our experience the experimental data fit the equation quite precisely and the constants can be used to correct reliably the panel meter readings P_i, to their correct value P_G, over extended periods of time.

3.3.3.2 Estimation of Line Loss

The standard power meter should now be used to measure the power levels P_A at the applicator input

for the same power settings P_i as used during measurements of P_G. In most instances the values of P_A are less than P_G for the same P_i, mainly due to line losses, as can be seen from Tables 3.8 and 3.9. In general

$$P_A = T \times P_G \tag{3.4}$$

where T is the line transmission factor having values between 0 and 1.

It is recommended that the factor T be measured separately for each power channel and for each set of line components used clinically. It should be recognized that the line loss is usually a greater problem at the MW frequencies than at the lower radiofrequencies.

3.3.3.3 Net Power at Applicator

By definition, the net power P_N at the applicator is the difference between the forward and reflected powers. However, if P_F and P_R are the readings of the

Table 3.8. Forward power (P_G) at generator output (BSD MG 202 generator)

Frequency (MHz)	Forward power (W)		% Deviation $[100 \times (P_i - P_m)/P_m]$
	Indicated (P_i)	Measured (P_m)	
241.1	11	7.5	46
241.1	50	44	13
241.1	101	90	12
241.1	158	131	13
241.1	198	180	10
241.1	248	229	8
241.1	303	270	10
241.1	349	311	12
241.1	398	359	11
241.1	412	370	11

Regression equation $P_G = 0.93 \ P_i$

Table 3.9. Forward power (P_A) at applicator input (BSD MG 202 generator)

Frequency (MHz)	Forward power (W)		% Deviation $[100 \times (P_i - P_m)/P_m]$
	Indicated (P_i)	Measured (P_m)	
247.8	10.5	4.9	114.3
247.8	49.5	32	54.7
247.8	98	64	53
247.8	150	98	52.5
247.8	202	134	50.3
247.8	249	170	46.5
247.8	306	203	50.7
247.8	354	231	53.4
247.8	397	262	51.3

Regression equations $P_A = 0.69 \ P_i$ and $P_A = 0.74 \ P_G$

forward and reflected powers on the instrument panels, because of the inherent errors due to meter offset, miscalibration, and line losses, P_N is not correctly represented by $(P_F - P_R)$. To derive P_N, first values of P_F and P_R must be corrected to "P_{FC}" and "P_{RC}" according to the equations

$$P_{FC} = A_F + B_F P_F + C_F P_F^2 \qquad (3.5)$$

and

$$P_R = A_R + B_R P_R + C_R P_R^2 , \qquad (3.6)$$

using the previously determined values of A, B, and C and then for line transmission effects. If T is the line transmission factor, just as the forward power is attenuated by a factor T in reaching the applicator, the power reflected at the applicator will also be attenuated by a factor T in returning to the meter before it is recorded. Hence, we estimate that the net power to the applicator is more correctly given by

$$P_N = T \times P_{FC} - P_{RC}/T . \qquad (3.7)$$

3.4 Applicator Evaluations

3.4.1 Characterization of Applicators by SAR Patterns

The applicator consists of an EM antenna, waveguide, or US transducer which delivers power over some surface area. The exact distribution of the power density depends on the design of the applicator. For hyperthermia applications, the heating ability of an applicator is best characterized by the SAR pattern it generates in a volume of tissue or tissue equivalent phantom. The measurement of SAR under dynamic conditions of conductive and convective heat losses is a difficult proposition. However, if a burst of power is applied for a very short duration Δt and the rise of temperature ΔT is measured so quickly thereafter that significant heat losses and temperature smearing cannot occur, then in principle the SAR = $C \Delta T/\Delta t$ where C is the specific heat of the medium. In practice, although these ideal assumption are often compromised because finite time periods are required to deposit sufficient power so as to induce a measurable temperature rise and also to record the data, valuable information of sufficient accuracy can still be obtained with proper precautions. The reproducibility of measured SAR distributions in addition to the time

factors can also be compromised if the generator frequency, phantom characteristics, or power coupling techniques are not exactly reproduced. Because of these factors the measurements of SAR to evaluate hyperthermia applicators are often difficult to reproduce and intercomparisons of data from two different measurements must be made with careful control of such extraneous factors.

The heating patterns in a phantom are most conveniently measured in 2-D with an infrared thermographic camera system. However, these rather expensive systems are not available in most clinics. Therefore, the QA intercomparisons of SAR must rely on selected data obtained with a few temperature probes embedded in a phantom. The temperature probes can be either permanently embedded or inserted within catheters placed at selected locations within the phantom. An advantage of catheters is that the probes can be moved either manually or by some semiautomated mechanism to obtain linear scans of SAR distributions.

In performing the QA tests for an applicator, three questions are of concern:

1. Are the SAR data supplied by the manufacturer reliable?
2. Is the SAR pattern affected by changes in patient geometry or coupling techniques?
3. Is the applicator performance consistent from treatment to treatment?

The QA tests, the frequency of testing, and the criteria in Table 3.7 are recommended to answer these questions. The rather broad criterion of 10% deviations is considered acceptable because in most clinical situations it is difficult to measure SAR distributions with much better accuracy and reproducibility.

3.4.2 Test Procedures

3.4.2.1 Check of Manufacturer-Supplied SAR Data

It is important that clinics insist on receiving SAR information for each applicator from the manufacturer and perform in-house QA tests to check the accuracy and reproducibility of this information. The minimum data provided by the manufacturer should include at least two orthogonal scans of relative SAR at some fixed depth and a scan along the central axis in the depth direction. Exact information on the frequency, phantom characteristics, and the coupling technique used for these measurements must also be provided.

The in-house QA test should exactly reproduce the material characteristics, phantom shape and size, EM frequency, and power coupling techniques used by the manufacturer. Also, unless a thermographic camera is available, nonperturbing thermometers must be used within judiciously placed catheters to ensure checking of the SAR values at the most clinically relevant positions. In order to minimize thermal smearing errors a short heating period of 30–60 s at a sufficiently high power setting and a fast scanning technique for recording temperature rise patterns must be used. The relative SAR measurement at any point should agree with the manufacturer's data to within 10%. If larger discrepancies are found, the manufacturer should be consulted to determine its cause.

3.4.2.2 SAR Patterns with Coupling Bolus

The manufacturer-supplied SAR data should not be used for clinical treatment planning without verifying their validity for actual treatment conditions. This is especially necessary if coupling bolus materials other than those used during phantom tests are used during patient treatments.

The SAR data for clinical use must be measured for the geometrical setups and coupling techniques actually used during patient treatments. At present there are no reliable criteria for determining clinical suitability of applicators based on the SAR data. A commonly quoted rule of thumb recommends that an applicator should not be considered suitable for treatment if its SAR pattern in a static phantom cannot enclose the entire tumor (or target) volume within the 50% isoSAR surface when the 100% SAR is defined as the maximum SAR anywhere in the medium.

3.4.2.3 Single Point SAR Reproducibility

In addition to the relative SAR patterns, the actual SAR at some fixed reference point, normalized to the net power into the applicator, is an important characteristic of any applicator. The reproducibility of the reference SAR from one day to another generates confidence that the heating system (i.e., the applicator, coupling devices, cables, and generator) is functioning in the normally expected manner.

The reference point can be judiciously chosen depending on the applicator and its applications. For external EM applicators used for superficial tumors, the reference point can be chosen to be at 1 cm depth along the central axis in a muscle-equivalent phantom.

The power-normalized SAR at the reference point is given by

$$(SAR)_{ref} = (\Delta T/\Delta t) \times C/P_{net} \, (W/kg \times W) \qquad (3.8)$$

where ΔT = rise in temperature, Δt = duration of heating, P_{net} = net power into applicator (see Sect. 3.3.3), and C = specific heat of the muscle phantom. A single permanent phantom with one thermometer embedded at the reference point can be used for a quick check of the reproducibility of $(SAR)_{ref}$. Although this test may not be possible or convenient for all applicators, we recommend that it be used whenever possible as a check of the heating system before each treatment.

3.4.3 Some Results of SAR Measurements

3.4.3.1 Effect of Phantom Size, Shape, and Composition

Although each applicator is in principle expected to have a characteristic SAR pattern, our investigations demonstrate that the measurements of SAR for the same applicator can be quite different depending on the measurement procedure. First we observe dramatic changes in SAR depending on the composition, size, and shape of the phantom used during measurements. Secondly, we find that drastic variations in the SAR patterns are caused by seemingly small changes in the coupling technique or efficiency.

The heating patterns in Figs. 3.11–3.15 were measured with an infrared thermographic camera (Agema 782, Infrared systems, Dandaryd, Sweden) attached to a digital computer system (800, BMC International, Osaka, Japan). The precision of digital temperature outputs from this system is estimated to be 0.1 °C. The data correspond to a 10×10 cm superficial dielectrically loaded waveguide applicator (Clini-Therm Corp., Dallas, Texas) operating at 915 MHz. A standard muscle equivalent gel (Guy et al. 1978; Chou et al. 1984) was used. It is composed of TX-150 (Oil Center Research, Lafayette, Ind.), 8.42% by weight; polyethylene powder, 15.44%; water, 75.15%; sodium chloride, 0.996%; and a preservative (Dowicil 75, Dow Chemical, Pittsburgh, California), 0.07%. The gel is enclosed in a 0.31 cm thick Plexiglas box that splits in the center. A polyester-silk screen separates the gel on each side of the split. Phantoms are allowed to equilibrate to room temperature before use for experiments.

In HPC's conventional (x, y, z) spatial reference system, the cross-section of an external EM applicator has x-direction perpendicular to the E-field and y-direction parallel to it. The z-direction represents the depth into the phantom; $x = 0$ is defined at the boundary of the muscle-equivalent material.

Two flat box-shaped gel phanta of different sizes, $14.5 \times 14.5 \times 6 \text{ cm}^3$ and $10.5 \times 10.5 \times 6 \text{ cm}^3$, and a curved cylindrical phantom 12 cm in diameter and 15 cm long were used to investigate the effect of phantom size and shape. Each phantom was heated for 30 s at a forward power of 100 W; reflected power was approximately $1-2$ W. The temperature rise patterns were obtained by digital subtraction of thermal images recorded under identical geometrical setup just "before" and "after" heating. The heating pattern is qualitatively seen in the Polaroid pictures of subtracted images in Figs. 3.11–3.13. In each figure, the top picture is the heat pattern on the surface of the phantom. The bottom picture is the heat pattern in depth direction (vertically downwards) in the principal plane perpendicular to the E-field direction for the applicator. The horizontal and vertical crossplots along the principal axes are also shown at the bottom and right of the image. The hottest regions are bright white, followed by yellow, red, violet, pale yellow, white, and blue in decreasing order of temperature rise.

Fig. 3.11, corresponding to the larger 14.5×14.5 cm phantom, shows the normally expected, central heating for the 10×10 cm applicator operating at 915 MHz. Fig. 3.12, however, shows a dramatic difference in the heating pattern when the smaller phantom ($10.5 \times 10.5 \text{ cm}^2$) is used with the $10 \times 10 \text{ cm}^2$ applicator. Similarly, Fig. 3.13 shows a completely different heating pattern compared to Fig. 3.11 when the

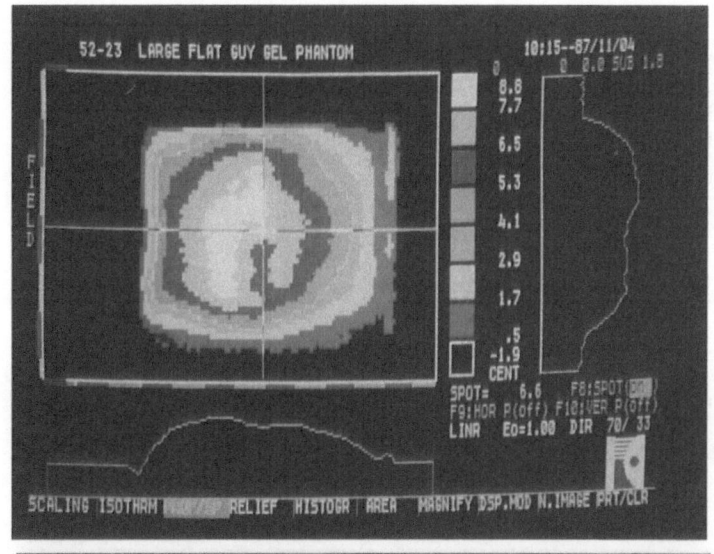

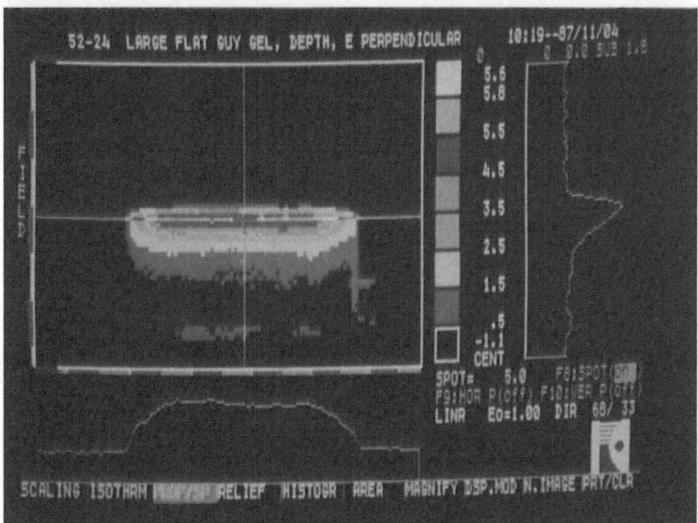

Fig. 3.11. a Central, symmetric surface heating and b exponentially decreasing heating at depth [$T = T_0 \text{Exp} (-2Z/Z_0)$, Z_0 approx. 3 cm] in a large $14.5 \times 14.5 \times 6 \text{ cm}^3$ phantom with a $10 \times 10 \text{ cm}^2$ 915-MHz external applicator

phantom is of cylindrical shape (simulating neck or thigh) rather than flat (simulating chest wall).

We have observed that the dramatic changes in the heating patterns due to small size or curved shape of the phanta can be reversed by using deionized water bolus bags to fill up the air spaces so as to reproduce the geometry of the original $14.5 \times 14.5 \, \text{cm}^2$ thick phantom used for obtaining the data in Fig. 3.11. This result points to the necessity of filling up the empty air spaces between the applicator and patient skin with bolus bags to achieve reproducibility of SAR patterns.

3.4.3.2 Effect of Phantom Composition

To investigate the difference in SAR patterns due to phantom composition, measurements in a phantom $10.5 \times 10.5 \times 6 \, \text{cm}^3$ composed of commercially available bologna were compared to those in a similar size gel phantom. We found basic heat patterns in the two phanta to be very similar although the rate of temperature rise on the surface in the bologna was $1.5 - 2$ times faster. This difference in SAR per watt is mainly due to the dielectric constant of the bologna.

Formulas for preparing muscle-equivalent phantom materials for a range of frequencies between 13.56 and 2450 MHz have been published (Chou et al. 1984). The dielectric constants and conductivity of these materials depend on temperature; hence it is important to keep the phantom at temperatures near $22 \, °\text{C}$ to simulate real tissue at $37 \, °\text{C}$. Also, the maximum temperature rise in phantom should be limited to $10 \, °\text{C}$ to avoid excessive errors in the SAR estimates. The useful lifetime of these phantoms is about

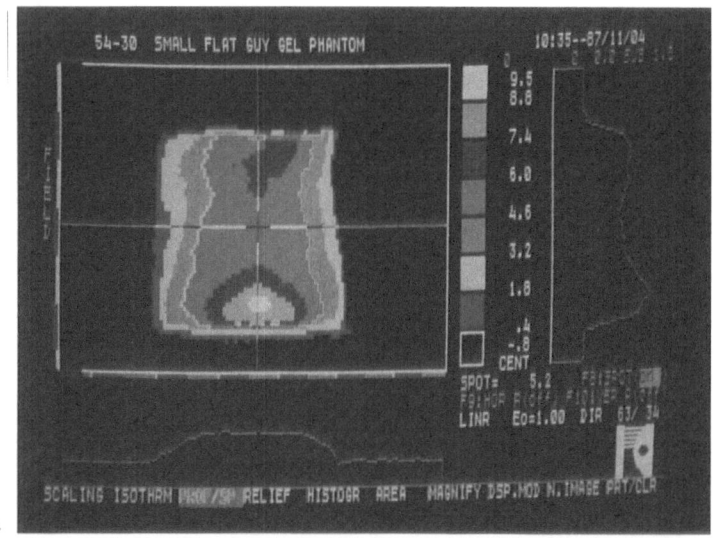

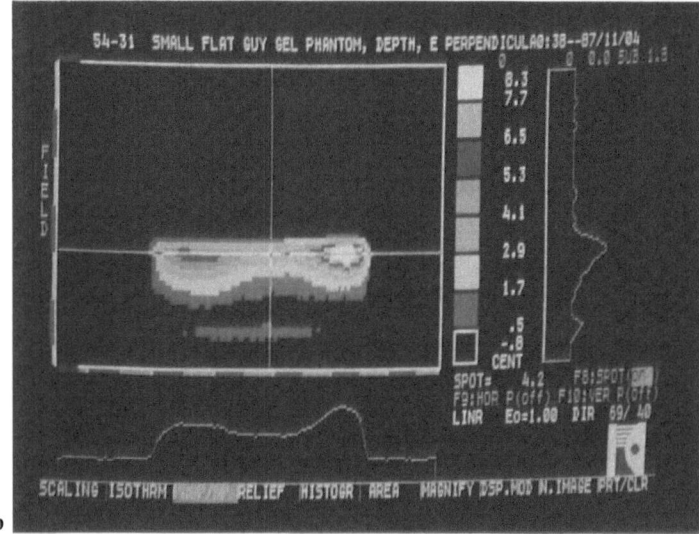

Fig. 3.12. a Distorted off-center surface heating and **b** peripheral hot spots in depth heating for the same applicator in a smaller $10.5 \times 10.5 \times 6 \, \text{cm}^3$ phantom

2–4 weeks, and it is important not to rely on old or deteriorated phantoms.

3.5 Evaluation of Coupling

3.5.1 Coupling Efficiency

The efficiency of power transfer from an applicator to the patient depends on the change of impedance at the interface. The efficiency often can be improved by inserting certain coupling materials, referred to as bolus, between the applicator and the patient's skin. Improved coupling efficiency results in increased forward power, decreased reflected power, and an overall reduction in power leakage. Improvements in coupling efficiency and reduction in leakage by proper selection of a bolus were demonstrated by Nussbaum et al. (1983) for an 8×8 cm^2 waveguide operating at 915 MHz in the TE$_{10}$ mode.

Our data on reflected power levels with three different types of applicator and different coupling mechanisms are shown in Table 3.10. A low ratio of reflected to forward power (P_R/P_F) indicates higher coupling efficiency. It should be noted that if the panel meter readings of P_R and P_F are used, the ratio P_R/P_F (column 4) is generally smaller than the ratio P_{RC}/P_{FC} (column 5) derived after the panel readings are corrected as described in Sect. 3.3.3.3. However, both columns indicate that the coupling efficiency is least with an open air gap between the applicator and the patient. Direct contact and the use of deionized water or

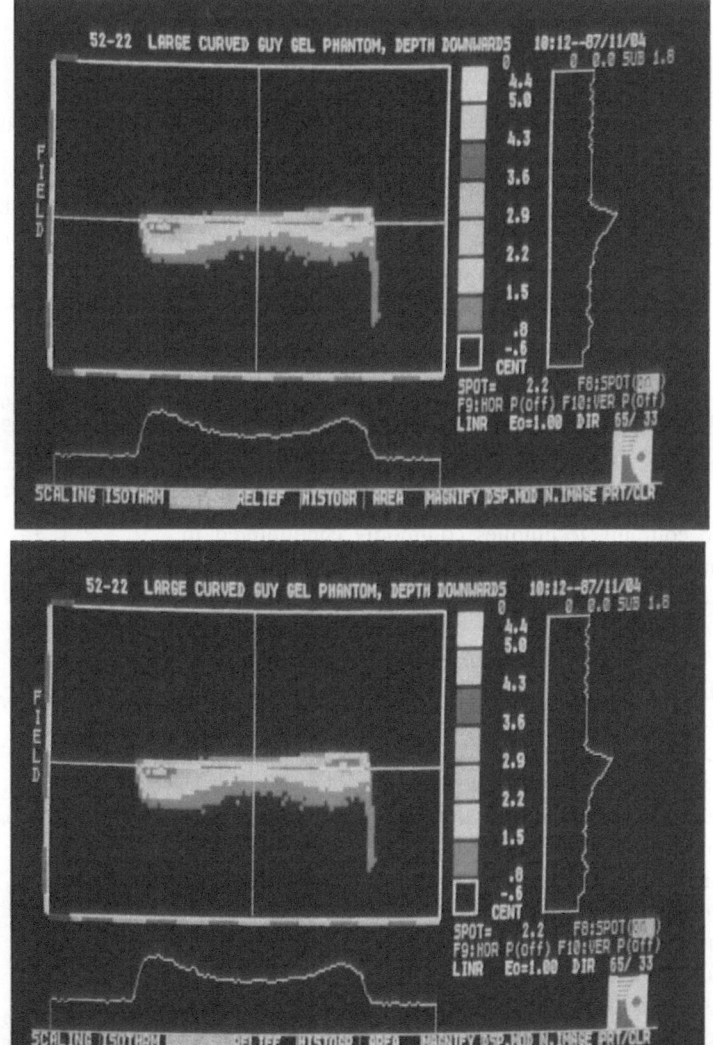

Fig. 3.13. a Noncentral surface heating perpendicular to E-field direction and **b** distorted heating at depth for the same applicator in a 12-cm diameter cylindrical phantom

Table 3.10. Effect of coupling bolus on reflected power

Applicators[a]	Coupling bolus	Planar phantom	$(P_{refl}/P_{forward})$	
			Meter indicators	Corrected[c]
CS-TEM-3	Air gap (1.1 cm)	P, muscle	52%	82%
CS-TEM-3	(Direct contact)	P, muscle	20%	32%
CS-TEM-3	(Direct contact)	1 cm fat, muscle	10%	16%
CTC 10×10	Water	P, muscle	3%	8%
CTC 10×10	Mineral oil	P, muscle	≥ 10%	≥ 27%
CTC 10×10	Mineral oil	Muscle	7%	18%
CTC 10×10	Water	Muscle	5%	14%
BSD MA-100	1.6 cm water	P, muscle	3%	6%
BSD MA-100	1.5 cm air gap	P, muscle	4%	7%
BSD MA-100	1.6 cm water	1 cm fat, muscle	7%	12%
BSD MA-200	2.5 cm air gap[b]	P, muscle	5%	10%

P, 0.31 cm Plexiglas
[a] CS-TEM-3 (Cheung Laboratories, Inc., Lanham-Seabrook, MD); CTC 10×10 (Clini-Therm Corp., Dallas, Texas); BSD MA-100 (BSD Medical, Salt Lake City, Utah)
[b] Water bags at ends to reduce stray radiation
[c] Corrected for power meter calibration and cable loss

mineral oil bolus improves the coupling efficiency for all three types of applicator.

In clinical practice it is important from a QA perspective to achieve correct placement and coupling of the applicator. During patient setup every attempt must be made to maximize coupling efficiency by minimizing reflected power and stray leakage.

3.5.2 Effect of Coupling Bolus on Heating Pattern

The coupling techniques commonly encountered in clinical applications of superficial EM applicators include direct contact, air gaps, water bolus, and mineral oil bolus.

A comparison of the SAR distributions in tissue-equivalent phantom when a 1.5-cm air gap or 1.6-cm deionized water bolus are used with a BSD MA-100 10×10 cm applicator (BSD Medical, Salt Lake City, Utah) is presented in Figs. 3.14 and 3.15. The curves are cross-plots of temperature rise patterns at depths from 1 to 4 cm in a direction perpendicular to the E-field. The drastic differences in the heat distributions due to change in bolus are clearly seen. This result emphasizes the need for evaluating the effect of coupling bolus on heating patterns on a case by case basis. Also, it suggests the importance of reproducing the coupling from session to session in a course of fractionated patient treatments.

3.6 Evaluation of Skin Cooling Devices

The skin cooling mechanisms are not yet standardized. They can consist of a simple air blower or cooling pads through which a precooled fluid is circulated. The QA concerns with these devices are whether or

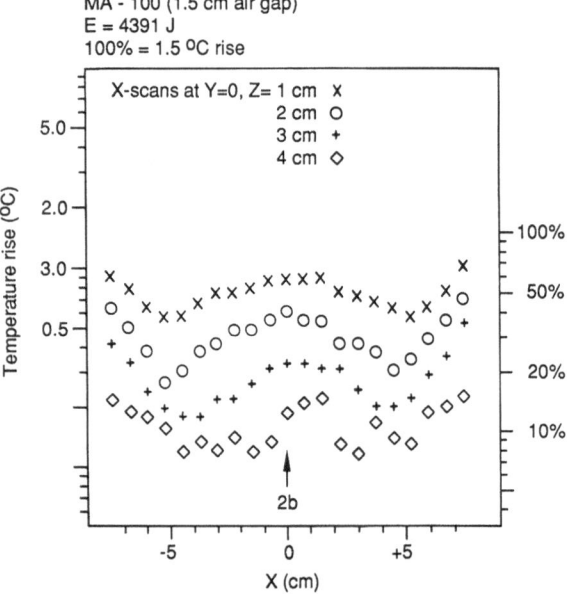

Fig. 3.14. Temperature rise pattern at depths of 1–4 cm in a direction perpendicular to the E-field for a 915-MHz applicator coupled with 1.5-cm air gap to a muscle-equivalent phantom

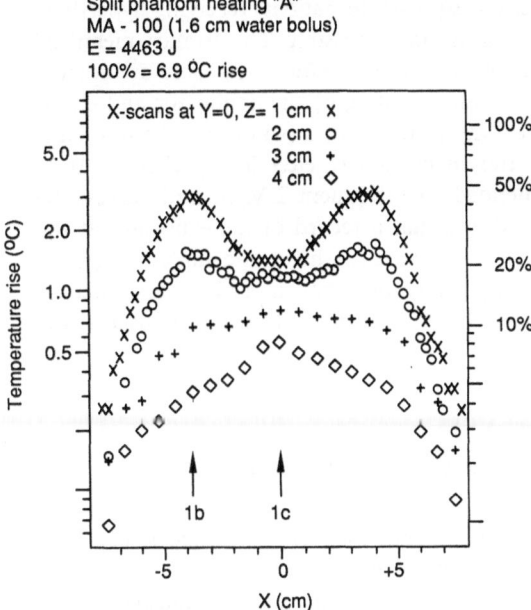

Split phantom heating "A"
MA - 100 (1.6 cm water bolus)
E = 4463 J
100% = 6.9 °C rise

Fig. 3.15. Temperature rise pattern at depths of 1–4 cm in a direction perpendicular to the E-field for a 915-MHz applicator coupled with a 1.6-cm deionized water bolus to a muscle-equivalent phantom

not they cool the requisite surface uniformly and whether they themselves absorb power during treatment. If power is absorbed, not only will the power coupling efficiency be reduced, but the cooling pad may become warmer than desirable. If the device has ridges and circulating fluids, it is also possible that they can distort the SAR patterns. The extent of the distortion can depend on the orientation of the ridges relative to the E-field direction and the fluid flow rate. QA checks of these devices, therefore, in addition to a simple evaluation of uniformity of cooling should include the evaluation of their influence on SAR patterns. For this purpose the cooling device must be included as part of the setup while measuring SAR patterns according to the procedures described in Sect.

3.5.2. If is also recommended that a temperature probe be used to monitor the real-time temperature of the cooling pads or fluid during treatment.

3.7 Evaluation of Electromagnetic Hazard

3.7.1 Operator Exposure to Electromagnetic Leakage Radiation

The safety standards recommended by the ANSI regarding human exposure to RF and MW fields, presented in Table 3.11, are used as guidelines for hyperthermia clinics in the United States. These recommendations are intended to apply to occupational as well as nonoccupational exposures but are not applicable to the intentional exposure of patients. So far there is no international consensus on these recommended safety limits.

In order to comply with the safety requirements, it is necessary that an EM leakage radiation survey meter be available at all times in every facility. The criteria for its calibration and the frequency of EM hazard surveys are indicated in Table 3.7. We suggest that the EM survey meter be calibrated when purchased and recalibrated yearly. Facilities using low radiofrequencies should have probes sensitive to both magnetic (H) and electric (E) fields. Surveys should be carried out with the applicator and the phantom setup to simulate patient treatments under various conditions. Locations showing leakage readings above those in column 2, Table 3.12, should be identified. QA procedures should be implemented to assure that operators are aware of the locations of high leakage above ANSI limits and are instructed to limit continuous self-exposure at these locations to less than 6 min.

Table 3.11. EM radiation safety recommendations[a]

Frequency range (MHz)	Power density[b,c] (mW/cm^2)	E^2 (V^2/m^2)	H^2 (A^2/m^2)
0.3 – 3	100	400000	2.5
3 – 30	900/f^2	4000 (900/f^2)	0.025 (900/f^2)
30 – 300	1.0	4000	0.025
300 – 1500	f/300	4000 (f/300)	0.025 (f/300)
1500 – 100000	5.0	20000	0.125

[a] "Safety levels with respect to human exposure to radiofrequency electromagnetic fields 300 kHz to 10 GHz," American National Standards Institute, Report C95-1, 1982, IEEE, Piscataway, NJ
[b] Continuous exposures over 6 min at power densities exceeding those in column 2 should be avoided by the operator
[c] f is frequency in MHz

3.7.2 Patient Exposure to EM Leakage Radiation

Although the ANSI limits are not applicable to intentional patient exposure, the unintentional and unnecessary exposure to patients (especially eyes, brain, gonads, and internal organs) should be minimized during treatment. Patient exposures to leakage radiation, especially from deep heating RF devices, can be far in excess of the ANSI limits (Shrivastava 1988). High exposures to patient organs are also possible with superficial MW heating devices. Although there is some disagreement among experts on the real hazards of such EM exposures and to date there are no guidelines to follow, clinical prudence dictates that

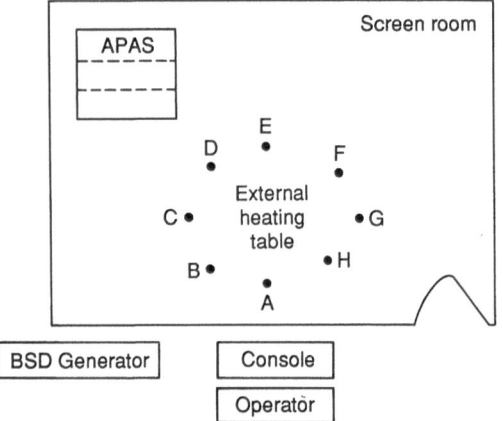

Fig. 3.16. Top view of a hyperthermia treatment room and locations of points A–H for EM leakage survey in Table 3.12

unnecessary exposure to patients receiving hyperthermia be kept as low as possible. The primary means of minimizing unnecessary patient exposure are correct placement and coupling of the applicator and elimination of any metal objects (wires, temperature electrodes, treatment tables, etc.) in its vicinity. It is recommended that frequent EM hazard surveys be conducted and that a record of the exposures at the position of the patient's head and abdomen (when not directly under the applicator) be maintained at least for each new type of treatment setup used at the facility.

3.7.3 Hazard Survey Procedure

We recommend that the stray EM-field strength be measured at a number of locations 1 m away from the applicator, at the site of the operator console, and at selected sites where high readings may be suspected, e.g., near cables, connectors, or metal objects in the room. The applicator itself should be set on a phantom in a geometry simulating actual treatment setup and the maximum power level used clinically should be used. The measurements in Table 3.12 were made with an isotropic broadband field strength meter (Model 3002, Holaday, Eden Prarie, Minnesota) with separate E- and H-field probes. The system was calibrated at the U.S. Food and Drug Administration. BSD MA-150 and MA-200 applicators operating at 147.5 to 320 MHz and coupled to a muscle-equivalent gel phantom were used. A top view of the room and

Table 3.12. Stray EM leakage; H-field reading

Applicator	MA-150		MA-200		MA-200	
Freq. (MHz)	242		147.5		320	
Power F (W)	131		224		200	
R (W)	8.2		0.7		14	
Reading	($\times 10^{-2}$ A²/M²)	mW/cm²	($\times 10^{-2}$ A²/M²)	mW/cm²	($\times 10^{-2}$ A²/M²)	mW/cm²
A	1	0.38	0	0	0.5	0.19
B	1	0.38	0	0	6	2.26
C	0.5	0.19	1	0.38	2.5	0.94
D	0.5	0.19	2.5	0.94	1	0.38
E	1	0.38	0.5	0.19	1.5	0.56
F	0.5	0.19	0.5	0.19	1	0.38
G	0.5	0.19	0.5	0.19	2	0.75
H	1	0.38	1	0.38	4	1.51[a]
I	0.5	0.19	1	0.38	1	0.38
J	1	0.38	5.5	2.08[a]	10	3.77

I = 1 m above applicator
J = 1 m below applicator
[a] Exceeds ANSI C95-1 standard of 1 mW/cm² for 30–300 MHz

Table 3.13. Maximum EM leakage levels observed on HPC survey

Generator/applicator	Frequency (MHz)	Forward/reflected power (W)	Maximum leakage (mW/cm^2)	Within acceptable criterion
BSD-1000/MA-150	421	115/15	0.42	Yes
BSD-1000/Single Horn	467	120/20	1.6	Yes
PWAL/S-8 Spiral	455	110/10	1.01	Yes
BSD-1000/APAS	60	1100/76	2.0	No
Henry/Magnetrode	13.56	980/0	18.9[a]	No
Induction coil	13.56	800/−	500[b]	No
BSD-1000/APAS	60	1050/46	1.32	No

[a] 1 m above applicator. Also 10.6 mW/cm^2 at 5 cm from operator console
[b] Near power cables between generator and applicator on treatment room floor

the locations of points (A−H) are shown in Fig. 3.16. Measured values of E- and H-field strengths are converted to plane wave-equivalent power densities by using a value of 377 ohms for the impedance of free space. At 320 MHz the ANSI recommendation (Table 3.11) requires leakage levels to be less than 1.06 mW/cm^2 at locations where continuous exposure to a person can exceed 6 min. Locations B, H and J exceed this limit.

3.7.4 Results of EM Leakage Surveys

The maximum leakage levels at distances >1 m for seven different hyperthermia units are presented in Table 3.13. The local MW heating devices (rows 1−3) met the ANSI standard for maximum permissible leakage but all the low frequency, deep or regional heating devices did not.

In our surveys we have found that every facility using a high power RF device for regional or deep heating had at least some locations potentially occupiable by an operator where the stray EM leakage exceeded the ANSI recommendation. Both the magnitude of EM leakage and the locations of high leakage areas are highly dependent on individual treatment conditions, phantom design, or coupling efficiency. Therefore, in-house QA procedures should entail frequent surveys and constant alertness on the part of operators to minimize prolonged self-exposure.

References

Chou CK, Chen GW, Guy A, Luk K (1984) Formulas for preparing phantom muscle tissue at various radiofrequencies. Bioelectromagnetics 5:435−441

Christensen DA (1983) Thermometry and thermography. In: Storm FK (ed) Hyperthermia in cancer therapy. Hall Medical, Boston, pp 223−232

Curley MG, Lele PP (1984) Some potential errors in measurement of temperatures in vivo during hyperthermia by ultrasound and electromagnetic energy. In: Overgaard J (ed) Hyperthermia oncology 1984, vol 1. Summary papers, proceedings of the 4th int symposium on hyperthermia oncology, Aarhus, Denmark, 2−6 July 1984. Taylor and Francis, London, pp 561−654

Dewey WC, Hopwood LE, Sapareto SA, Gerweck LE (1977) Cellular responses to combinations of hyperthermia and radiation. Radiology 123:463−479

Fessenden P, Lee ER, Samulski TV (1984) Direct temperature measurements. Cancer Res 4 (Suppl):4799−4804

Guy AW, Lehmann JF, Stonebridge JB, Sorensen CC (1978) Development of a 915-MHz direct contact applicator for therapeutic heating of tissue. IEEE Trans Microwave Theory Tech MTT 26:550−556

Lee ER, Fessenden P (1984) Evaluation of parylene insulated, flexible multiple junction thermocouple temperature probes for ultrasound hyperthermia. Proceedings of the 5th annual meeting, North American Hyperthermia Group, Orlando

Nussbaum GH, Goodman RA, Bruce AA (1983) Improved applicator-patient coupling in microwave-induced hyperthermia. Med Phys 10(5):897−898

Saylor TK, Shrivastava PN, Paliwal BR (1984) Performance and QA tests for thermometers used in hyperthermia (abstract). Radiation Research Society Abstracts, 16. Academic, San Diego

Saylor TK, Matloubieh AY, Shrivastava PN (1988) Quality assurance for thermometry. In: Paliwal B, Hetzel F, Dewirst M (eds) Biological, physical and clinical aspects of hyperthermia. AAPM Monograph 16. American Institute of Physics, New York NY, pp 380−395

Shrivastava PN, Saylor TK, Matloubieh AY, Paliwal BR (1988a) Hyperthermia thermometry evaluation: criteria and guidelines. Int J Radiat Oncol Biol Phys 14:327−335

Shrivastava PN, Saylor TK, Matloubieh AY, Paliwal BR (1988b) Hyperthermia quality assurance results. Int J Hyperthermia 4:25−37

Shrivastava PN, Luk K, Oleson J, Dewhirst M, Pajak T, Paliwal B, Perez C, Sapareto S, Saylor T, Steeves R (1989) Hyperthermia Quality Assurance Guidelines. Int J Radiat Oncol Biol Phys 16:571−587

Subject Index